KU-378-169

MATHS ON THE BACK OF AN ENVELOPE

ROB EASTAWAY

MATHS ON THE BACK OF AN ENVELOPE

CLEVER WAYS TO (ROUGHLY) CALCULATE ANYTHING

HarperCollins*Publishers*

South Tyneside Libraries	
JAR	
1071894 X	
Askews & Holts	23-Sep-2019
510	£9.99

www.harpercollins.co.uk

First published by HarperCollins*Publishers* 2019

1 3 5 7 9 10 8 6 4 2

© Rob Eastaway 2019

Rob Eastaway asserts the moral right to be identified
as the author of this work.

A catalogue record of this book is available from the British Library

ISBN 978-0-00-832458-2

Printed and bound in Great Britain by CPI Group (UK) Ltd, Croydon

All rights reserved. No part of this publication may be reproduced,
stored in a retrieval system, or transmitted, in any form or by any means,
electronic, mechanical, photocopying, recording or otherwise, without
the prior written permission of the publishers.

MIX
Paper from
responsible sources
FSC™ C007454

FSC
www.fsc.org

This book is produced from independently certified FSC™ paper
to ensure responsible forest management.

For more information visit: www.harpercollins.co.uk/green

CONTENTS

PROLOGUE

HOW MANY CATS?

A few years ago, at a school event, I asked the audience of teenagers to submit some estimation questions that I would then attempt to answer live on stage. One pupil posed this simple question: 'How many cats are there in the world?'

Cats are always a popular topic, so I took it on.

My thinking went like this:

Let's assume that most cats are domestic.

Some people have more than one cat, but usually a household has only *one* cat, if any at all.

In the UK, and thinking of my own street as an example, it seems reasonable to suppose that there might be one cat in every five households.

And, if a household contains on average two people, that means there is one cat for every 10 people.

So, with 70 million people in the UK, let's say that there are, perhaps, seven million cats in the UK.

So far, so good. But what about the number of cats in the rest of the world? It seems unlikely that cats are as popular in countries like India or China as they are in the UK (although what would I know? Remember, this is purely guesswork on my part), therefore, I'd expect the ratio across the world to be smaller than it is in the UK – maybe one cat for every 20 people?

So, with eight billion people in the world, that suggests there are maybe:

$$8 \ billion \div 20 = 400 \ million \ cats.$$

It doesn't seem an outrageous number.

That was the figure I suggested, anyway.

A member of the audience put his hand up.

'The real number is 600 million,' he said.

'Really – how do you know?'

'I just looked it up online.'

So that's it: no need to come up with an answer; it's already been done.

If it's really that simple, we can forget about doing estimations altogether. With just a few clicks on Google you will probably find a statistic to answer every question you could possibly think of.

Except for one crucial thing.

Where did the person who published that figure of 600 million online get *their* number from? No one, I'm quite sure, has gone around the world doing a census of cats. The figure of 600 million is an estimate. It might be an estimate that is based on a slightly more scientific method than mine, using rigorous surveys and cross-checks, but it's just as likely that the figure cited online came from somebody who did a back-of-envelope calculation like the one I just described. Or, perhaps they simply invented a figure that suited their agenda. There's no reason to believe that their number is more reliable than mine – indeed, it might be less reliable.

Once a statistic like this is out there, published in a newspaper, quoted on a website, it becomes 'fact', and it can be re-quoted so often that the source may go unquestioned and very quickly be forgotten altogether.

It's an important reminder that the majority of statistics published anywhere are estimates, many of them worked out on the equivalent of the back of an envelope. When back-of-envelope calculations produce a very different answer from the one that's been put forward, it doesn't necessarily mean that the estimate is wrong. It means rather that the published figure deserves *more scrutiny*.

We tend to think of maths as being an 'exact' discipline, where answers are right or wrong. And it's true that there is a huge part of maths that is about exactness.

But in everyday life, numerical answers are sometimes just the start of the debate. If we are trained to believe that every numerical question has a definitive, 'right' answer then we miss the fact that numbers in the real world are a lot fuzzier than pure maths might suggest.

I've realised in writing this book that there is a kind of paradox. On the one hand, I want to argue that *approximate* numbers can often be more informative, and more trustworthy, than precise numbers. Yet at the same time, in order to be able to produce those approximate answers, it's essential to know how to do some calculations *exactly*. Your basic times tables, for example. Concrete, exact maths is the foundation for the woolly numbers that we have to deal with in everyday life.

I've divided the book into four sections.

In the first section, I explore how precise numbers can be misleading, and why it's good not to be entirely dependent on a calculator.

The second section includes the arithmetical techniques and the other knowledge that is an essential foundation if you want to embark on back-of-envelope calculations. This includes a refresher on how to do arithmetic that you may not have needed to practise since you left primary school, as well as short cuts that you probably never encountered there.

The rest of the book shows how to use these techniques to tackle problems, from everyday conversions, to more serious issues like helping the environment. And at the end, there is a collection of so-called Fermi questions: quirky and esoteric challenges to come up with a reasonable answer based on very little hard data.

Back-of-envelope maths is an important and valuable life skill. But that's not its only benefit. Many of us also indulge in it simply because it's a fun and stimulating exercise that keeps the brain sharp.

1

THE PERILS OF PRECISION

ENVELOPES VERSUS CALCULATORS

I don't know when the backs of envelopes first became a popular place for jotting rough-and-ready calculations. Was it before or after people started to use 'the back of a fag packet', or in the USA, the back of a napkin?

Regardless of where the expression was first coined, the back of an envelope[1] has come to symbolise any sort of rough-and-ready type of calculation that gives an indication of what the right answer will be.

It is the tool that people in business use for quickly checking the viability of a new project. Engineers use it to check if a proposed solution is likely to work. And commentators on statistics use it to help make sense of the myriad numbers that are thrown out by politicians, 'expert' pundits and marketers.

---

[1] You don't literally need to use the back of an envelope of course; any scrap of paper will do. Which is just as well, because while I was writing this book, envelopes were removed from the list of essential items used in calculating the Retail Price Index, a sign that they are no longer a standard household item.

On a more mundane level, it's the maths you might use every day to ensure you aren't getting ripped off by a so-called 'deal' that turns out to be anything but.

It is also maths and arithmetic that can be done without needing to resort to a calculator.

But wait a minute. *Maths without a calculator?* To many people, this notion seems quaintly old-fashioned, or even masochistic. Why grapple with manual or mental calculations when most of us have a phone (with a calculator) readily to hand almost all of the time?

This is not an anti-calculator book. Calculators are indispensable tools that have enabled us to do in seconds what used to take minutes, hours or even days. If you need to know exactly what £31.40 × 96 is, then unless you are a savant or somebody with plenty of time on your hands, a calculator is the only sensible option for working it out. And I'm probably typical in usually having a calculator – or a spreadsheet – to hand if I'm doing my tax return, or totting up expenses after a work-related trip.

But much of the time we don't need to know the exact answer. It's an approximate figure that matters. The point of back-of-envelope maths is to help see the bigger picture behind numbers.

Suppose a sales team has a target of £10,000. If they report that they have sold 96 units at £31.40 each – that's roughly:

100 × £30 = £3,000 revenue.

That's massively short of the £10,000 target, even if the estimate is out by a few per cent.

When the government announces a £1 billion increase in health spending, is that significant? Spread between 50 million people? It won't be exactly one billion pounds of course, nor will it be spread evenly between 50 million people, but with back-of-envelope maths, we can work out it will represent an average of something nearer to £20 (i.e. hardly anything) than £200 per person.

Of course, even these simplified calculations can be done on a calculator. But the reality is that they rarely are.

The argument: 'Who needs to do arithmetic when we all have calculators?' is usually a red herring. In situations where a calculation is not essential, most of us do it in our heads or on the back of an envelope, or don't do it at all.

And there are some who use their ability to figure things mentally to their advantage. I have a friend who made his fortune as a wheeler-dealer in finance. I asked him to share some advice.

'I have two tips for succeeding when negotiating a deal with somebody,' he said. 'The first is: learn how to be able to read upside down, so that you can decipher the documents of the person opposite you. And my second is: be able to do the calculations faster than they can.'

TEST YOURSELF

How is your arithmetic without the aid of a calculator? Try these 10 questions. There is no time pressure, and you're allowed to use pencil and paper if you want. As you do these questions, you might want to think about *how* you do them. Are you recalling facts you've memorised? Do you use a pencil-and-paper method?

(a) 17 + 8
(b) 62 − 13
(c) 2,020 − 1,998
(d) 9 × 4
(e) 8 × 7
(f) 40 × 30
(g) 3.2 × 5
(h) One-quarter of 120
(i) What is 75% as a fraction?
(j) What is 10% of 94?

Find out how other people do these calculations on page 185.

I can still remember the thrill when I first got a calculator of my own. It was made by Commodore, and had red LED digits and buttons that made a satisfying click when you pressed them. It was a Christmas present, and I was 16 years old. I was captivated. Just being able to enter a number like 123456 and press the square root button was enough to send a tingle of excitement down my spine, as I gazed at all those digits after the decimal point. I'd never seen numbers to such precision before.

There were two things that came out of the arrival of cheap calculators.

The first was that we could all now do calculations that we would never have conceived of doing before. It was empowering, liberating and gave us a chance to see the bigger picture of mathematics without getting bogged down in the nitty-gritty of calculation.[2]

The second thing that happened was that we could now quote answers to several decimal places. The square root of 83? Certainly, sir, just give me one second – and how many digits would you like after the decimal point?

What could possibly go wrong?

---

[2] For example, what interesting answer do you get if you multiply $3 \times 7 \times 11 \times 13 \times 37$? Without a calculator, only somebody with strong mental calculation skills, an excess of curiosity, a lot of perseverance and a shortage of other things to do, would bother to find out. Even with a calculator in your pocket, you are still wondering if it's worth the effort. (Go on, you know you want to.)

SPURIOUS PRECISION

A tourist in a natural history museum was very impressed by the skeleton of a Tyrannosaurus Rex.

'How old is that fossil?' she asked one of the guides.

'It's 69 million years and 22 days,' said the guide.

'That's incredible, how do you know the age so precisely?' asked the tourist.

'Well, it was 69 million years old when I started working at the museum, and that was 22 days ago,' replied the guide.

The thoughtless precision of the museum guide in this old joke nicely illustrates why there is no point in stating a number to several figures if the overall measurement is only a rough estimate. Yet it is a mistake that is made time and again when presenting and interpreting numbers in everyday life.

Quoting a number to more precision than is justified is often called spurious accuracy, though it should really be called spurious *precision* and we will encounter it several times in this book. It is one of the strongest arguments against the unthinking overuse of calculators. The fact that you *can* work out numbers to several decimal places at the touch of a button doesn't mean that you *should*.

PRECISION VERSUS ACCURACY

The words precision and accuracy are often used interchangeably, to indicate how 'right' a measurement or number is. It is certainly possible for a number to be accurate and precise; for example: 74 × 23.2 = 1,716.8.

But used mathematically, precision and accuracy mean different things.

'Accuracy' is an indication of how close you are to the right answer. Suppose we are playing darts. I throw a dart at a dartboard and just miss the bullseye. My throw was quite accurate, but if you then throw and hit the bullseye, your throw was more accurate than mine. Likewise . . . if I tot up the items in my shopping basket and estimate that the total will be £65 while you reckon it will be £70, and the bill turns out to be £69.43, then you were more accurate than I was.

*Precision, on the other hand, is an indication of how confident you are in a number to a particular level of detail, so that you or somebody else would come up with the same figure if you did a measurement or calculation again. If you think the shopping basket will add up to £69, you are confident that you are right to the nearest pound; but if you suggest the bill will be £69.40, you are being more precise, and are confident your figure is right to the nearest 10 pence. Even more precise is £69.41. In maths terms, precision is about how many **significant figures** (this is an important concept – please see page 179) you can quote a number to.*

As a society, we put a lot of faith in precision. If we see a number such as 84.36, we tend to believe that the person who produced the number is confident of that figure to the second decimal place. We might even honour them with the label of 'expert' because they were able to produce a number to such precision. But people who produce 'precise' figures often abuse our trust, and accidentally or deliberately they imply a level of confidence that is not justified. When we read that 59,723 people attended an Arsenal football match, we are being led to believe that an exact tally was taken as fans went through the turnstiles. So when we discover that the actual number in the ground was closer to 50,000, we feel like we have been conned.[3]

When it comes to the use of numbers to interpret the world, accuracy is more important than precision. After all, a measurement that is accurate but not precise can still be helpful. But a measurement that is precise but inaccurate is not just unhelpful, it can sometimes be dangerous.

One of the unintended consequences of calculators is that they will give answers to as many decimal places as will fit on the screen – and in doing so, they tempt us to work to a level of precision that is often not justified.

DRESSING UP NUMBERS WITH DRESSAGE

In 2012 London hosted the Olympics. People throughout Britain celebrated as gold medallists stepped up on the podium in all sorts of sports in which previous GB athletes had rarely excelled.

---

[3] Discrepancies like this are common, because the attendances reported by clubs are typically the number of tickets sold, including debenture and season tickets, rather than the number of bodies that turned up on the day.

There was particular joy when Charlotte Dujardin, who had worked her way up from being a stable girl to becoming an elite equestrian, claimed Britain's first ever gold medal with her horse Valegro in the dressage event.

Dujardin's winning score from the judges was an impressive 90.089%.

Sportspeople often talk about 'giving it that extra ten per cent', but in this case it seemed as if Dujardin had fine-tuned this, so she could put in that extra 10.001%. What was it that made her performance better than somebody who got, say, 90.088% instead?

To understand where her three-decimal-place score came from, we need to look at how the judges allocated marks in that competition.

In the dressage event at the London Olympics, the competitors were required to take their horse through a series of movements, which would be assessed by seven judges seated around the arena so they could view from different angles. The judges were scoring under 21 headings: 16 of them were 'technical' marks given for how well specific movements were carried out, and five of them 'artistic', applying to the overall performance, with descriptions such as 'Rhythm, energy and elasticity' and 'Harmony between horse and rider' (yes, really).

Each item was marked out of 10, with half-marks allowed, but some scores were then given more weighting, and the five artistic scores were all multiplied by four. In total, each judge could give up to:

240 technical marks + 160 artistic merit marks = 400 marks total.

It meant each rider had a total of $7 \times 400 = 2,800$ points to play for.

There is an element of subjectivity when assessing how good a horse's performance has been, so it is hardly surprising that the judges don't all give the same mark. For a particular technical movement, one judge might give it an '8', while another spots a slightly dropped shoulder and reckons it's a '7'. In fact, in Dujardin's case, the judges' total marks ranged from 355 to 370, and when added together she got a total of 2,522.5 points out of a maximum possible 2,800.

And this is where the percentage comes in, because her 2,522.5 total was then divided by 2,800 to give a score out of one hundred, a percentage:

$$2,522.5 \div 2,800 = 90.089\%.[4]$$

Well, actually that's not the exact number. It was really 90.089285714285714 ... %.

Indeed, this number never stops, the pattern 285714 repeats for ever. This is what happens when you divide a number by a multiple of 7. So Dujardin's score had to be rounded, and the authorities who were responsible for the scoring system decided to round scores to three decimal places.

What would have happened if Dujardin had been awarded half a mark less by the judges? She would have scored:

$$2,522 \div 2,800 = 90.071\%.$$

[4] In the 2016 Olympics in Rio, Charlotte Dujardin surpassed her London score, achieving an impressive 93.857% to claim gold once again.

In other words, the precision of her actual score of 90.089 was misleading. It wasn't possible to score any other mark between 90.089% and 90.071%, Dujardin didn't give it an extra 0.001%, but rather she gave it that extra 0.018%. Quoting her score to two decimal places (i.e. 90.09%) was enough.

The second decimal place is needed to guarantee that two contestants with different marks don't end up with the same percentage, but it still gives a misleading sense of the accuracy of the scoring. In reality, each judge 'measures' the same performance differently. A half-mark disagreement in the artistic score (which is then multiplied by 4, remember) shifts the overall mark by 0.072%. And the actual discrepancies between the judges were bigger than that. For 'Harmony between horse and rider' one judge marked her 8 out of 10 while another gave her 9.5 out of 10.

A NUMBER IS ONLY AS STRONG AS ITS WEAKEST LINK

There's a time and a place for quoting numbers to several decimal places, but dressage, and other sports in which the marking is subjective, isn't one of them.

By using this scoring system, the judges were leaving us to assume that we were witnessing scoring of a precision equivalent to measuring a bookshelf to the nearest millimetre. Yet the tool they were using to measure that metaphorical bookshelf was a ruler that measured in 10-centimetre intervals. And it was worse than that, because it's almost as if the judges each had different rulers and, on another day, that very same performance might have scored anywhere between, say, 89% and 92%. It was a score with potential for a lot of variability – more of which in the next section.

All of this reveals an important principle when looking at statistical measurements of any type. In the same way that a chain is only as strong as its weakest link, a statistic is only as reliable as its most unreliable component. That dinosaur skeleton's age of 69 million years and 22 days was made up of two components: one was accurate to the nearest million years, the other to the nearest day. Needless to say, the 22 days are irrelevant.

BODY TEMPERATURE A BIT LOW?
BLAME IT ON SPURIOUS PRECISION

In 1871, a German physician by the name of Carl Reinhold Wunderlich published a ground-breaking report on his research into human body temperature. The main finding that he wanted to publicise was that the average person's body temperature is 98.6 degrees Fahrenheit, though this figure will vary quite a bit from person to person.

The figure of 98.6 °F has become gospel,[5] the benchmark body temperature that parents have used ever since when checking if an unwell child has some sort of fever.

Except it turns out that Wunderlich didn't publish the figure 98.6 °F. He was working in Celsius, and the figure he published was

---

[5] In the USA, 98.6 °F is still the figure commonly used. In the UK, the commonly used figure in my parents' generation was 98.4 °F. Fahrenheit has now dropped out of fashion in the UK, and doctors have reverted to 37 °C as the norm.

37 °C, a rounded number, which he qualified by saying that it can vary by up to half a degree, depending on the individual and on where the temperature is taken (armpit or, ahem, orifice).

The figure 98.6 came from the translation of Wunderlich's report into English. At the time, Fahrenheit was the commonly used scale in Britain. To convert 37 °C to Fahrenheit, you multiply by 9, divide by 5 and add 32; i.e. 37 °C converts to 98.6 °F. So the English translation – which reached a far bigger audience than the German original, gave the figure 98.6 °F as the human norm. Technically, they were right to do this, but the decimal place created a misleading impression. If Wunderlich had quoted the temperature as 37.0 °C, it would have been reasonable to quote this as 98.6 °F, but Wunderlich deliberately didn't quote his rough figure to the decimal place. For a figure that can vary by nearly a whole degree between healthy individuals, 98.6 °F was (and is) spurious precision. And in any case, a study in 2015 using modern, more accurate thermometers, found that we've been getting it wrong all these years, and that the average human temperature is 98.2 °F, not 98.6 °F.

VARIABILITY

In the General Election of May 2017, there was a shock result in London's Kensington constituency. The sitting MP was a Conservative with a healthy majority, but in the small hours of the Friday, news came through that the result was too close to call, and there was going to be a recount. Hours later, it was announced that there needed to be a *second* recount. And then, when even that failed to resolve the result, the staff were given a few hours to get some sleep, and then returned for a *third* recount the following day.

Finally, the returning officer was able to confirm the result: Labour's Emma Dent Coad had defeated Victoria Borwick of the Conservatives.

The margin, however, was tiny. Coad won by just 20 votes, with 16,333 to Borwick's 16,313.

You might expect that if there is one number of which we can be certain, down to the very last digit, it is the number we get when we have counted something.

Yet the truth is that even something as basic as counting the number of votes is prone to error. The person doing the counting might inadvertently pick up two voting slips that are stuck together. Or when they are getting tired, they might make a slip and count 28, 29, 40, 41 . . . Or they might reject a voting slip that another counter would have accepted, because they reckon that marks have been made against more than one candidate.

As a rule of thumb, some election officials reckon that manual counts can only be relied on within a margin of about 1 in 5,000 (or 0.02%), so with a vote like the one in

Kensington, the result of one count might vary by as many as 10 votes when you do a recount.[6]

And while each recount will typically produce a slightly different result, there is no guarantee which of these counts is actually the correct figure – if there is a correct figure at all. (In the famously tight US Election of 2000, the result in Florida came down to a ruling on whether voting cards that hadn't been fully punched through, and had a hanging 'chad', counted as legitimate votes or not.)

Re-counting typically stops when it is becoming clear that the error in the count isn't big enough to affect the result, so the tighter the result, the more recounts there will be. There have twice been UK General Election votes that have had seven recounts, both of them in the 1960s, when the final result was a majority below 10.

All this shows that when it is announced that a candidate such as Coad has received 16,333 votes, it should really be expressed as something vaguer: 'Almost certainly in the range 16,328 to 16,338' (or in shorthand, $16,333 \pm 5$).

If we can't even trust something as easy to nail down as the number of votes made on physical slips of paper, what hope is there for accurately counting other things that are more fluid?

In 2018, the two Carolina states in the USA were hit by Hurricane Florence, a massive storm that deposited as much as 50 inches of rain in some places. Among the chaos,

---

[6] According to Susan Loynes, an Electoral manager who worked on the count, the Kensington recounts were all within five of each other, and the variation was entirely down to judgements about spoiled ballot papers. Given the small variation, the Conservatives were clutching at straws when they requested the third recount.

a vast number of homes lost power for several days. On 18 September, CNN gave this update:

> 511,000—this was the number of customers without power Monday morning—according to the US Energy Information Administration. Of those, 486,000 were in North Carolina, 15,000 in South Carolina and 15,000 in Virginia. By late Monday, however, the number [of customers without power] in North Carolina had dropped to 342,884.

For most of that short report, numbers were being quoted in thousands. But suddenly, at the end, we were told that the number without power had dropped to 342,884. Even if that number were true, it could only have been true for a period of a few seconds when the figures were collated, because the number of customers without power was changing constantly.

And even the 486,000 figure that was quoted for North Carolina on the Monday morning was a little suspicious – here we had a number being quoted to three significant figures, while the two other states were being quoted as 15,000 – both of which looked suspiciously like they'd been rounded to the nearest 5,000. This is confirmed if you add up the numbers: 15,000 + 15,000 + 486,000 = 516,000, which is 5,000 higher than the total of 511,000 quoted at the start of the story.

So when quoting these figures, there is a choice. They should either be given as a range ('somewhere between 300,000 and 350,000') or they should be brutally rounded to just a single significant figure and the qualifying word 'roughly' (so, 'roughly 500,000'). This makes it clear that these are not definitive numbers that could be reproduced if there was a recount.

And, indeed, there are times when even saying 'roughly' isn't enough.

Every month, the Office for National Statistics publishes the latest UK unemployment figures. Of course this is always newsworthy – a move up or down in unemployment is a good indicator of how the economy is doing, and everyone can relate to it. In September 2018, the Office announced that UK unemployment had fallen by 55,000 from the previous month to 1,360,000.

The problem, however, is that the figures published aren't very reliable – and the ONS know this. When they announced those unemployment figures in 2018, they also added the detail that they had 95% confidence that this figure was correct to *within 69,000*. In other words, unemployment had fallen by 55,000 plus or minus 69,000. This means unemployment might actually have gone down by as many as 124,000, or it might have gone *up* by as many as 14,000. And, of course, if the latter turned out to be the correct figure, it would have been a completely different news story.

When the margin of error is larger than the figure you are quoting, there's barely any justification in quoting the statistic at all, let alone to more than one significant figure. The best they can say is: 'Unemployment probably fell slightly last month, perhaps by about 50,000.'

It's another example of how a rounded, less precise figure often gives a fairer impression of the true situation than a precise figure would.

SENSITIVITY

We've already seen that the statistics should really carry an indication of how much of a margin of error we should attach to them.

An understanding of the margins of error is even more important when it comes to making predictions and forecasts.

Many of the numbers quoted in the news are predictions: house prices next year, tomorrow's rainfall, the Chancellor's forecast of economic growth, the number of people who will be travelling by train . . . all of these are numbers that have come from somebody feeding numbers into a spreadsheet (or something more advanced) to represent this mathematically, in what is usually known as a *mathematical model* of the future.

In any model like this, there will be 'inputs' (such as prices, number of customers) and 'outputs' that are the things you want to predict (profits, for example).

But sometimes a small change in one input variable can have a surprisingly large effect on the number that comes out at the far end.

The link between the price of something and the profit it makes is a good example of this.

Imagine that last year you ran a face-painting stall for three hours at a fair. You paid £50 for the hire of the stand, but the cost of materials was almost zero. You charged £5 to paint a face, and you can paint a face in 15 minutes, so you did 12 faces in your three hours, and made:

£60 income − £50 costs = £10 profit.

There was a long queue last year and you were unable to meet the demand, so this year you increase your charge from £5 to £6. That's an increase of 20%. Your revenue this year is £6 × 12 = £72, and your profit climbs to:

$$£72 \ income - £50 \ costs = £22 \ profit.$$

So, a 20% increase in price means that your profit has more than doubled. In other words, your profit is extremely sensitive to the price. Small percentage increases in the price lead to much larger percentage increases in the profit.

It's a simplistic example, but it shows that increasing one thing by 10% doesn't mean that everything else increases by 10% as a result.[7]

EXPONENTIAL GROWTH

There are some situations when a small change in the value assigned to one of the 'inputs' has an effect that grows dramatically as time elapses.

Take chickenpox, for example. It's an unpleasant disease but rarely a dangerous one so long as you get it when you are young. Most children catch chickenpox at some point unless they have been vaccinated against it, because it is highly infectious. A child infected with chickenpox might

---

[7] There are some markets that can be *extremely* sensitive. A friend who worked in the retail (petrol station) side of one of the major oil companies reckoned that if they priced petrol at, say, 132.9p per litre, while a competitor next door was charging 0.1p less at 132.8p, that difference of less than 0.1% in the price could lose them at least 5% of their custom.

typically pass it on to 10 other children during the contagious phase, and those newly infected children might themselves infect 10 more children, meaning there are now 100 cases. If those hundred infected children pass it on to 10 children each, within weeks the original child has infected 1,000 others.

In their early stages, infections spread 'exponentially'. There is some sophisticated maths that is used to model this, but to illustrate the point let's pretend that in its early stages, chickenpox just spreads in discrete batches of 10 infections passed on at the end of each week. In other words:

$$N = 10^T,$$

where N is the number of people infected and T is the number of infection periods (weeks) so far.

After one week: $N = 10^1 = 10$.
After two weeks: $N = 10^2 = 100$.
After three weeks: $N = 10^3 = 1,000,$
and so on.

What if we increase the rate of infection by 20% to N = 12, so that now each child infects 12 others instead of 10? (Such an increase might happen if children are in bigger classes in school or have more playdates, for example.)

After one week, the number of children infected is 12 rather than 10, just a 20% increase. However, after three weeks, $N = 12^3 = 1,728$, which is heading towards double what it was for N = 10 at this stage. And this margin continues to grow as time goes on.

CLIMATE CHANGE AND COMPLEXITY

Sometimes the relationship between the numbers you feed into a model and the forecasts that come out are not so direct. There are many situations where the factors involved are inter-connected and extremely complex.

Climate change is perhaps the most important of these. Across the world, there are scientists attempting to model the impact that rising temperatures will have on sea levels, climate, harvests and animal populations. There is an overwhelming consensus that (unless human behaviour changes) global temperatures will rise, but the mathematical models produce a wide range of possible outcomes depending on how you set the assumptions. Despite overall warming, winters in some countries might become colder. Harvests may increase or decrease. The overall impact could be relatively benign or catastrophic. We can guess, we can use our judgement, but we can't be certain.

In 1952, the science-fiction author Raymond Bradbury wrote a short story called 'A Sound of Thunder' in which a time-traveller transported back to the time of the dinosaurs accidentally kills a tiny butterfly, and this apparently innocuous incident has knock-on effects that turn out to have changed the modern world they return to. A couple of decades later, the mathematician Edward Lorenz is thought to have been referencing this story when he coined the phrase 'the butterfly effect' as a way to describe the unpredictable and potentially massive impact that small changes in the starting situation can have on what follows.

These butterfly effects are everywhere, and they make

confident long-term predictions of any kind of climate change (including political and economic climate) extremely difficult.

MAD COWS AND MAD FORECASTS

In 1995, Stephen Churchill, a 19-year-old from Wiltshire, became the first person to die from Variant Creutzfeldt–Jakob disease (or vCJD). This horrific illness, a rapidly progressing degeneration of the brain, was related to BSE, more commonly known as 'Mad Cow Disease', and caused by eating contaminated beef.

As more victims of vCJD emerged over the following months, health scientists began to make forecasts about how big this epidemic would become. At a minimum, they reckoned there would be at least 100 victims. But, at worst, they predicted as many as 500,000 might die – a number of truly nightmare proportions.[8]

Nearly 25 years on, we are now able to see how the forecasters did. The good news is that their prediction was right – the number of victims was indeed between 100 and 500,000. But this is hardly surprising, given how far apart the goalposts were.

The actual number believed to have died from vCJD is about 250, towards the very bottom end of the forecasts, and about 2,000 times smaller than the upper bound of the prediction.

But why was the predicted range so massive? The reason is that, when the disease was first identified, scientists could make a

---

[8] Five hundred thousand was the worst-case scenario in a forecast by the European Union Scientific Steering Committee in 1996.

reasonable guess as to how many people might have eaten contaminated burgers, but they had no idea what proportion of the public was vulnerable to the damaged proteins (known as prions). Nor did they know how long the incubation period was. The worst-case scenario was that the disease would ultimately affect everyone exposed to it – and that we hadn't seen the full effect because it might be 10 years before the first symptoms appeared. The reality turned out to be that most people were resistant, even if they were carrying the damaged prion.

It's an interesting case study in how statistical forecasts are only as good as their weakest input. You might know certain details precisely (such as the number of cows diagnosed with BSE), but if the rate of infection could be anywhere between 0.01% and 100%, your predictions will be no more accurate than that factor of 10,000.

At least nobody (that I'm aware of) attempted to predict a number of victims to more than one significant figure. Even a prediction of '370,000' would have implied a degree of accuracy that was wholly unjustified by the data.

DOES THIS NUMBER MAKE SENSE?

One of the most important skills that back-of-envelope maths can give you is the ability to answer the question: 'Does this number make sense?' In this case, the back of the envelope and the calculator can operate in harmony: the calculator does the donkey work in producing a numerical answer, and the back of the envelope is used to check that the number makes logical sense, and wasn't the result of, say, a slip of the finger and pressing the wrong button.

We are inundated with numbers all the time; in particular, financial calculations, offers, and statistics that are being used to influence our opinions or decisions. The assumption is that we will take these figures at face value, and to a large extent we have to. A politician arguing the case for closing a hospital isn't going to pause while a journalist works through the numbers, though I would be pleased if more journalists were prepared to do this.

Often it is only after the event that the spurious nature of a statistic emerges.

In 2010, the Conservative Party were in opposition, and wanted to highlight social inequalities that had been created by the policies of the Labour government then in power. In a report called 'Labour's Two Nations', they claimed that in Britain's most deprived areas '54% of girls are likely to fall pregnant before the age of 18'. Perhaps this figure was allowed to slip through because the Conservative policy makers wanted it to be true: if half of the girls on these housing estates really were getting pregnant before leaving school, it painted what they felt was a shocking picture of social breakdown in inner-city Britain.

The truth turned out to be far less dramatic. Somebody had stuck the decimal point in the wrong place. Elsewhere in the report, the correct statistic was quoted, that 54.32 out of every 1,000 women aged 15 to 17 in the 10 most deprived areas had fallen pregnant. Fifty-four out of 1,000 is 5.4%, not 54%. Perhaps it was the spurious precision of the 54.32' figure that had confused the report writers.

Other questionable numbers require a little more thought. The National Survey of Sexual Attitudes has been published every 10 years since 1990. It gives an overview of sexual behaviour across Britain.

One statistic that often draws attention when the report is published is the number of sexual partners that the average man and woman has had in their lifetime.

The figures in the first three reports were as follows:

Average (mean) number of opposite-sex partners in lifetime (ages 16–44)

	Men	Women
1990–91	8.6	3.7
1999–2001	12.6	6.5
2010–2012	11.7	7.7

The figures appear quite revealing, with a surge in the number of partners during the 1990s, while the early 2000s saw a slight decline for men and an increase for women.

But there is something odd about these numbers. When sexual activity happens between two opposite-sex people, the overall 'tally' for all men and women increases by one. Some people will be far more promiscuous than others, but across the whole population, it is an incontravertible fact of life that the total number of male partners for women will be the same as the number of women partners for men. In other words, the two averages ought to be the same.

There are ways you can attempt to explain the difference. For example, perhaps the survey is not truly representative – maybe there is a large group of men who have sex with a small group of women that are not covered in the survey.

However, there is a more likely explanation, which is that somebody is lying. The researchers are relying on individuals' honesty – and memory – to get these statistics, with no way of checking if the numbers are right.

What appears to be happening is that either men are exaggerating, or women are understating, their experience. Possibly both. Or it might just be that the experience was more memorable for the men than for the women. But whatever the explanation, we have some authentic-looking numbers here that under scrutiny don't add up.

THE CASE FOR
BACK-OF-ENVELOPE THINKING

I hope this opening section has demonstrated why, in many situations, quoting a number to more than one or two significant figures is misleading, and can even lull us into a false sense of certainty. Why? Because a number quoted to that precision implies that it is accurate; in other words, that the 'true' answer will be very close to that. Calculators and spreadsheets have taken much of the pain out of calculation, but they have also created the illusion that any numerical problem has an answer that can be quoted to several decimal places.

There are, of course, situations where it is important to know a number to more than three significant figures. Here are a few of them:

- In financial accounts and reports. If a company has made a profit of £2,407,884, there will be some people for whom that £884 at the end is important.
- When trying to detect small changes. Astronomers looking to see if a remote object in the sky has shifted in orbit might find useful information in the tenth significant figure, or even more.
- Similarly in the high end of physics there are quantities linked to the atom that are known to at least 10 significant figures.
- For precision measurements such as those involved in GPS, which is identifying the location of your car or your destination, and where the fifth significant figure might

mean the difference between pulling up outside your friend's house and driving into a pond.

But take a look at the numbers quoted in the news – they might be in a government announcement, a sports report or a business forecast – and you'll find remarkably few numbers where there is any value in knowing them to four or more significant figures.

And if we're mainly dealing with numbers with so few significant figures, the calculations we need to make to find those numbers are going to be simpler. So simple, indeed, that we ought to be able to do most of them on the back of an envelope or even, with practice, in our heads.

2

TOOLS OF THE TRADE

THE ESSENTIAL TOOLS OF ESTIMATION

For most back-of-envelope calculations, the tools of the trade are quite basic.

The first vital tool is the ability to round numbers to one or more significant figures.

The next three tools are ones that require exact answers:

- Basic arithmetic (which is built around mental addition, subtraction and a reasonable fluency with times tables up to 10).
- The ability to work with percentages and fractions.
- Calculating using powers of 10 (10, 100, 1,000 and so on) and hence being able to work out 'orders of magnitude'; in other words, knowing if the answer is going to be in the hundreds, thousands or millions, for example.

And finally, it is handy to have at your fingertips a few key number facts, such as distances and populations, that crop up in many common calculations.

This section will arm you with a few tips that will help you with your back-of-envelope calculations later on – including a technique you may not have come across that I call Zequals, and how to use it.

ARE YOU AN ARITHMETICIAN?

In the opening section there was a quick arithmetic warm-up. It was a chance to find out to what extent you are an *Arithmetician*.

Arithmetician is not a word you hear very often.

In past centuries it was a much more familiar term. Here, for example, is a line from Shakespeare's *Othello*: 'Forsooth, a great arithmetician, one Michael Cassio, a Florentine.' That line is spoken by Iago, the villain of the play, who is angry that he has been passed over for the job of lieutenant by a man called Cassio. It is an amusing coincidence that Shakespeare's arithmetician Cassio has a name very similar to Casio, the UK's leading brand of electronic calculator.

Iago scoffs that Cassio might be good with numbers, but he has no practical understanding of the real world. (This rather harsh stereotype of mathematical people as being abstract thinkers who are out of touch with reality is one that lives on today.)

Shakespeare never used the word 'mathematician' in any of his plays, though in Tudor times the two words were often used interchangeably, just as 'maths' and 'arithmetic' are today – much to the annoyance of many mathematicians.

So what is the difference between maths and arithmetic? If you ask mathematicians this question, they come up with many different answers. Things like 'being able to logically prove what is true' and 'seeing patterns and connections'. What they never say is: 'knowing your times tables' or 'adding up the bill'.

Arithmetic, on the other hand, is entirely about calculations.

Here's an example to show what I mean:

Pick any whole number (789, say). Now double it and add one. By using a logical proof, a mathematician can say with absolute certainty that the answer will be an odd number, even if they are unable to work out the answer to 'what is twice 789 add one?'[9]

On the other hand, an arithmetician can quickly and competently work out that $(789 \times 2) + 1 = 1,579$, without needing a calculator.

The strongest arithmeticians can do much harder calculations, too. They can quickly work out in their head what $4/7$ is as a percentage; can multiply 43×29 to get the exact answer; and can quickly figure out that in a limited-overs cricket match, if England require 171 runs in 31 overs they'll need to score at a bit more than five and a half runs per over.

My mother, who left school at 17, was a strong arithmetician, as were many in her generation. That was almost inevitable. A large part of her schooling had been daily practice filling notebooks with page after page of arithmetical exercises. But she knew little about algebra, geometry

[9] Though, to be fair, most mathematicians *would* be able to work that out.

or doing a formal proof, in the same way that many top mathematicians are hopeless at arithmetic.

There is, however, a huge amount of overlap between arithmetic and mathematics. Many arithmetical techniques and short cuts lead on to deep mathematical ideas, and most of the maths that is studied up until school-leaving age requires an element of arithmetic, even if it's no more than basic multiplication and addition. Arithmetic and maths are both grounded in logical thinking, and both exploit the ability (and joy) of seeing patterns and connections.

And yet, although arithmetic crops up everywhere, after the age of 16 it is very rarely studied. Almost without exception, public exams beyond 16 allow the use of a calculator, and most people's arithmetical skills inevitably waste away after GCSE.

A while ago, a friend who runs an engineering company was talking with some final-year engineering undergraduates about a design problem he was working on. 'We have this pipe that has a cross-sectional area of 4.2 square metres,' he said, 'and the water is flowing through at about 2 metres per second, so how much water is flowing through the pipe per second?' In other words, he was asking them what 4.2×2 equals. He was assuming that these bright, numerate students would come back instantly with '8.4' or (since this was only a rough-and-ready estimate) 'about 8'. To his dismay, all of them took out their calculators.

Calculators have removed the need for us to do difficult arithmetic. And it's certainly not essential for you to be a strong arithmetician to be able to make good estimates. But it helps.

TEST YOURSELF

Can you quickly estimate the answer to each of these 10 calculations? If you get within (say) 5% of the right answer, you are already a decent estimator. And if you are able to work out exactly the right answers to most of them *in your head*, that's a bonus, and you can call yourself an arithmetician.

(a) A meal costs £7.23. You pay £10 in cash. How much change do you get?

(b) Mahatma Gandhi was born in October 1869 and died in January 1948. On his last birthday, how old was he?

(c) A newsagent sells 800 chocolate bars at 70p each. What are his takings?

(d) Kate's salary is £28,000. Her company gives her a 3% pay rise. What is her new salary?

(e) You drive 144 miles and use 4.5 gallons of petrol. What is your petrol consumption in miles per gallon?

(f) Three customers get a restaurant bill for £86.40. How much does each customer owe?

(g) What is 16% of 25?

(h) In an exam you get 38 marks out of a possible 70. What is that, to the nearest whole percentage?

(i) Calculate 678 × 9.

(j) What is the square root of 810,005 (to the nearest whole number)?

Answers on page 188.

BASIC ARITHMETIC

ADDITION AND SUBTRACTION

The classic written methods for arithmetic start at the right-hand (usually the units) column and work to the left. But when it comes to the sort of speedy calculations that are part of back-of-envelope thinking, it generally pays to work from the left instead.

For example, take the sum: 349 + 257.

You were probably taught to work it out starting from the units column at the right. The first step would be:

$$
\begin{array}{r}
3\ \ 4\ \ 9 \\
+\ \ 2\ \ 5_1\ 7 \\
\hline
6
\end{array}
$$

9 + 7 = 16, write down the 6 and 'carry' the 1.[10]

You then continue working leftwards:

$$
\begin{array}{r}
3\ \ 4\ \ 9 \\
+\ \ 2_1\ 5_1\ 7 \\
\hline
6\ \ 0\ \ 6
\end{array}
$$

4 + 5 + 1 = 10, write down the 0 and 'carry' the 1; 3 + 2 + 1 = 6.

[10] The '1' being carried represents 10. Where you place that little 1 depends on where you went to school and what the standard practice was at the time.

Working this out mentally, however, it is generally more helpful to start with the most significant digits (i.e. the ones on the left) first.

So the calculation 349 + 257 starts with 300 + 200 = 500, then add 40 + 50 = 90, and finally add 7 + 9 = 16. The advantage of working from the left is that the very first step gives you a reasonable estimate of what the answer is going to be ('it's going to be 500 or so . . .').

A similar idea applies to subtraction. Using the standard written method, working from the right, 742 − 258 requires some 'borrowing' (maybe you used different language). Here's the method my children learned at school:

$$
\begin{array}{r}
{}^{6}\!\!\not7 \ {}^{3}\!\!1\!\!\not4 \ 12 \\
- \ 2 \ 5 \ 8 \\
\hline
4 \ 8 \ 4
\end{array}
$$

8 from 2 can't be done, borrow 10,
12 − 8 = 4,
5 from 3 can't be done, borrow 10,
13 − 5 = 8, 2 from 6 = 4.

Starting from the left, however, you can read it as 700 − 200 (= 500), then 40 − 50 (so subtract 10 from 500) and, if you want the exact answer, calculate the units 2 − 8 (subtract 6).

MULTIPLICATION AND TIMES TABLES

Calculators may be here to stay, but children are still expected to learn their times tables in primary school, just as they were one hundred years ago.

In the UK, this means learning all multiplications up to 12. In some countries, such as India, it's not uncommon for this to be pushed to 20, so that some children might, for example, learn the answer to 13 × 17 off by heart.

When it comes to back-of-envelope maths, knowing your tables up to 10 is generally enough.

You may be a little rusty on your times tables. I'm guessing that the very fact that you are reading this means you can probably calculate 3 × 4 in your head, but many adults are out of practice when it comes to some of the harder multiplications. Notorious for tripping people up is 7 × 8, though according to one analysis of over a million calculations using times table that were done online,[11] it is 9 × 3 that is answered incorrectly the most often.

Here are a few tips that can be handy when doing multiplications in your head. These apply to the times tables, but are also handy for multiplying larger numbers.

Tip 1
The order of multiplication makes no difference to the answer. For example, 3 × 5 is the same as 5 × 3. One way to convince yourself why this is true is to think of multiplication as counting eggs in a tray.

[11] The *Times Tables Rockstars* website, created by Bruno Reddy, records pupils' performances every time they practise their times tables. There are over one billion items in their database, most of them still not analysed.

How many eggs are in the tray above? Three rows of five, or five columns of three? Either way, it comes to 15. What's powerful about this truism is that you can think of any multiplication as being like counting eggs in a tray. It means you can be confident that $7,431 \times 278$ is the same as $278 \times 7,431$, even if you don't know what the answer is.

This idea is not just restricted to multiplying two numbers together: $5 \times 13 \times 2$ is the same as $2 \times 5 \times 13$. By rearranging the order of the numbers that you are multiplying, you can often make a calculation easier. In this case, since $2 \times 5 = 10$, we can arrange $5 \times 13 \times 2$ to become $10 \times 13 = 130$.

Tip 2
Multiplying by 3 is the same as doubling a number, then adding the number again. Thus, 3×12 is the same as 2×12 (24), then add another 12.

Tip 3
Multiplying by four is the same as doubling, then doubling again. And to multiply a number by eight, you double it three times.

Tip 4
Instead of multiplying by nine, you can simply multiply your starting number by 10 and then subtract the starting number. For example, 9×8 is the same as 10×8 (= 80), minus 8 (= 72). Likewise, 9×68 is the same as 10×68 (680) minus 68 (= 612).

Tip 5

Multiplying by 5 is the same as halving the answer and then multiplying by 10. So, 468 × 5 looks hard. But it is the same as 468 ÷ 2 (= 234) × 10 (= 2,340), which is considerably easier. You can, of course, reverse the order and multiply by 10, then divide by two: 43 × 5 = 430 ÷ 2 = 115.

TEST YOURSELF

Trying working these out in your head (the short cuts mentioned above might help, or use your own method):

(a) 3 × 26
(b) 35 × 9
(c) 4 × 171
(d) 5 × 462
(e) 1,414 ÷ 5

Answers on page 190.

DIVISION

Division can be described in many ways, but one way to think of it is simply as the reverse of multiplication . . . working your times tables backwards. To divide 72 by 8, you mentally check what multiple of 8 gives 72 in the times table (answer 9). More often than not, there will be a remainder, but the idea is the same. So, 74 ÷ 8? The nearest multiple of 8 below 74 is 9 × 8 = 72, so the answer is 9 remainder 2. That's another reason why fluency with times tables is useful.

Dividing into larger numbers can be done using short division, which is just repeated working out using tables. To work out 596 divided by 4, the script I was taught goes like this:

$$1 \ 4 \ 9$$
$$4 \ \overline{)5 \ ^1 9 \ ^3 6}$$

Four into five goes once, remainder 1 (write 1 on the top; 1 is carried to make 19), four into 19 goes four times remainder 3, four into 36 goes exactly nine times.

You may wonder where this precise method belongs in a book about estimation. The point is that there is no need to follow the calculation through to the end – you can round the answer at any stage. For example, we could have stopped after the second division to get the answer 150 (rounded to two significant figures). Short division is a useful aid for calculating percentages, as we'll see on page 49.

DECIMALS AND FRACTIONS

PLACE VALUE AND DECIMAL POINTS

For some people, numbers become more difficult when they are smaller than one.

Thousands	Hundreds	Tens	Units		Tenths	Hundredths	Thousandths
6	*7*	*1*	*5*	.	*4*	*3*	*8*

The numbers to the right of the decimal point work in just the same way as those to the left. The first digit after the decimal point is the number of 'tenths', the next is 'hundredths', then 'thousandths' and so on.

In the number 0.528 there are five tenths and two hundredths, but you are allowed to read across the columns if you want, so you can also say there are 52.8 hundredths. Another way to write this is 52.8 ÷ 100, or 52.8 'per cent'. There's more about per cents below.

There may be times in everyday life when you need to convert fractions to decimals (or to percentages). A newspaper item might say: '1 in 4 have suffered some sort of theft, while 8% have experienced a burglary'. That's 25% + 8% = 33%: about one-third.

For many fractions, the decimal equivalent is familiar:

½ is the same as 0.5
¼ is 0.25
⅓ is 0.33

But what about two-sevenths (which is $2 \div 7$)?

One way to convert a fraction to a decimal is with short division – in exactly the same way as you'd work out $200 \div 7$, but with a decimal point inserted, since you'll be dealing with numbers smaller than 1:

$$0.2\ 8\ 5\ 7...$$
$$7\ \overline{)\ 2.^2 0\ ^6 0\ ^4 0\ ^5 0}$$

As a decimal, two-sevenths begins 0.2857 . . . which you can round to 0.286, or 0.29 etc., depending on the level of precision that you want.

DECIMAL POINTS – A MATTER OF LIFE OR DEATH?

A child weighing 20 kg has an infection and needs a course of the anti-biotic amoxicillin. The guideline is to administer 25 mg per kg of body weight of amoxicillin every 12 hours. The medication comes in a sus-pension of 250 mg per 5 ml. What dose (in ml) should the child be given?

That might sound like the GCSE maths question from hell, but it is in fact a fairly routine problem that might be encountered by a house doctor or senior nurse working on a hospital ward. Have a go at work-ing out the answer, and then imagine what it's like committing yourself to writing down the dose, knowing that if you put the decimal point in the wrong place, the consequences could be life-threatening.

A calculator might help here, of course, but only if you know which numbers to divide into which other ones – and be careful that fat fingers don't press the wrong digit, or an extra zero by mistake.

It shouldn't be a surprise that GPs and hospital workers do sometimes make mistakes with calculations like this. A doctor told me (on promise of being kept anonymous) that on one occasion, he prescribed a drug to a patient, and noticed that a couple of days later the patient's condition had, if anything, got slightly worse. Wondering why the drug wasn't working, he checked back and discovered he'd got the dose wrong by a factor of 10. Fortunately, he was giving the patient 10 times too little.

MULTIPLYING FRACTIONS

It's not often that you'll have a need to multiply fractions together. By far the most common reason I ever have to do this is when working out probabilities of two events happening (for example, what's the chance that the next card turned over in poker will be a Queen and the one after that will be another Queen?).

The simple rule for multiplying two fractions is to multiply the two top numbers (numerators) together to make the new numerator, and do the same with the bottom numbers (denominators) to make the new denominator.

For example:

$$\frac{2}{3} \times \frac{5}{13} = \frac{10}{39}$$

You can simplify the calculation if any of the numbers at the top and bottom of the fractions have a 'factor' (i.e. a number that divides into them) in common. For example, this . . .

$$\frac{6}{7} \times \frac{4}{15}$$

. . . looks difficult. But the 6 on top and the 15 on the bottom are both divisible by 3, so we can simplify it to:

$$\frac{6^{\,2}}{7} \times \frac{4}{15^{\,5}} = \frac{8}{35}$$

How big is $^8/_{35}$? Well, $8 \div 32$ is a quarter, so $8 \div 35$ is a bit less than a quarter.

TEST YOURSELF

(a) $\dfrac{1}{3} \times \dfrac{1}{2}$

(b) $\dfrac{2}{5} \times \dfrac{1}{4}$

(c) $\dfrac{3}{4} \times \dfrac{1}{5} \times \dfrac{2}{3}$

(d) Is $\dfrac{6}{7} \times \dfrac{14}{23}$ bigger or smaller than $\dfrac{1}{2}$?

(e) Work out $\dfrac{51}{52} \times \dfrac{50}{51}$

Answers on page 190.

PERCENTAGES

It's handy to remember that 'per cent' simply means 'divided by 100' and the 'of' in 'percentage *of* . . .' can be translated as 'and multiplied by . . .' In other words, 30% of 40 is the same as '30 divided by 100, and then multiplied by 40'.

This means that in any calculation in the style 'Find A per cent of B' (for example, 'Find 30% of 40'), the answer is going to be A times B divided by 100.

30% of 40?
30 × 40 = 1,200 . . . divide by 100 . . . 12%.

9% of 80?
9 × 80 = 720 . . . divide by 100 . . . 7.2%.

Here are some other tips for working out percentages:

Tip 1
Working out 10% of something is easy, so use that as a base. For example, what is 5% of 320? Ten per cent (one-tenth) of 320 is 32, so 5% will be half of that, which is 16.

Tip 2
If 10% doesn't get you quickly to your answer, try using 1% instead, and multiply up from there. For example, what is 3% of 80? One per cent of 80 is 0.8, so 3% is three times that, which is $3 \times 0.8 = 2.4$.

Tip 3

In calculations requiring you to work out the 'percentage of . . .', you can switch the order of the numbers, just as you can in any multiplication. Sixteen per cent of 25 is the same as 25% of 16. And 25% of 16 is the same as saying 'one-quarter of 16'. In other words, 16% of 25 = 4.

Tip 4

If you are confident with short division (see page 43), you can quickly become adept at working out percentages to two significant figures in your head – which is as precise as you are ever likely to need a percentage estimate.[12]

For example, if you score 57 out of 80 in a test, what's that as a percentage? We can work it out as follows:

$$57 \div 80 = 5.7 \div 8.$$

Now do this as short division:

$$\begin{array}{r} 0.712\ldots \\ 8\overline{\smash{\big)}\,5.7^10^20} \end{array}$$

So, it's 0.712, which is 71%, to two significant figures (or 'roughly 70%').

---

[12] In most news stories, percentages are quoted to two significant figures, and even when they are quoted to three, two significant figures is normally enough. For example, a story in front of me now talks about recycling rates of 44.8%. If the story had said 45% instead, would we have felt any less well informed?

TEST YOURSELF

(a) A shirt is marked as costing £28, but the shop is offering '25% off all marked prices'. What will the new price be?

(b) Work out 15% of 80.

(c) What is 14% of 50?

(d) Estimate 49 out of 68 as a percentage.

(e) Work out 266 ÷ 600 as a percentage to two significant figures.

(f) Kate's salary is £25,000 and she gets a promotion and an 8.4% pay rise. What is her new salary?

Answers on page 191.

REMOVING VAT

A percentage calculation that notoriously catches people out is removing the VAT from a price. At the time of writing, VAT in the UK is 20%. If a price is advertised as being £30 + VAT, then the total price can be worked out either by calculating 20% of £30 (= £6) and adding it to get £36 – or, more directly, by multiplying the original price by 1.20:

$$£30 \; (excl. \; VAT) \times 1.20 = £36 \; (inc. \; VAT).$$

So if the price of something is £36 including VAT, does that mean we can remove the VAT simply by taking off 20%? No! Twenty per cent of £36 is £7.20, which means the price without VAT would be:

$$£36 \text{ (inc. VAT)} - £7.20 = £28.80.$$

This is wrong! We know that the price without VAT was £30. What has happened?

To work out the price without VAT, you need to reverse what you did when you added VAT. To add VAT you multiply by 1.2, so to remove VAT you divide by 1.2:

$$£36 \text{ (inc. VAT)} \div 1.2 = £30 \text{ (excl. VAT)}.$$

Incidentally, dividing by 1.2 is equivalent to multiplying by $\frac{5}{6}$, and a quick way to work out the VAT element in a price that includes VAT (at 20%) is to divide the price by six. So, if the price of an item is £66 including VAT, the VAT element is £66 ÷ 6 = £11. The price without VAT is 5 × £11 = £55. (When VAT was 17.5% during the 1990s, there was no such simple short cut!)

CALCULATING WITH POWERS OF TEN

MULTIPLICATION

Knowing 7×8 is one thing, but what about 70×80, or $7,000 \times 800$? Many back-of-envelope calculations involve numbers in the hundreds, thousands and beyond, so it's important to be able to manipulate these numbers with ease.

Can you work out 700×80 without a calculator? A mental short cut that I have always used for a calculation with whole numbers such as 700×80 is to treat the leading digits and the zeroes separately. First multiply 7×8 ($= 56$), and then count up the zeroes at the end of the two numbers (there are three altogether) and stick them on the end. So the answer is 56,000.

I've asked hundreds of British teenagers to work out 700×80, though typically I posed it as a money question: 'A newsagent sells 700 chocolate bars at 80p each. What is the shop's total revenue?'

The reassuring thing is that the vast majority of teenagers know that $7 \times 8 = 56$, several years after their drilling in primary school. Where many struggle is in knowing how many zeroes to put into the answer, and where to stick the decimal point. In answer to the chocolate bar challenge, a significant minority will answer £56 or £5.60 or £5,600, or £56,000. And if teenagers struggle with this (including those who have already passed GCSE maths) it's safe to assume that many adults do too.

By the way, this is just the sort of problem where estimation can help: 80p is close to £1, and $700 \times £1 = £700$, so the correct answer is going to be a bit less than £700, and it certainly won't be £56 or £5,600.

TEST YOURSELF

(a) 400 × 90
(b) 300 × 700
(c) 80,000 × 1,100
(d) Bristol Old Vic Theatre needed to raise money for a complete refurbishment of the building. To help fund-raising, they created 50 'silver' tickets priced at £50,000 each, that would give the purchaser the right to every performance in the theatre in perpetuity. How far did this go towards their ultimate target of £25 million?

Answers on page 191.

Multiplication of decimals can be more fiddly. If one of the numbers has zeroes and the other is a decimal, you can 'trade' zeroes from one number to the other to make the calculation more manageable. In other words, you multiply one of the numbers by 10 and divide the other by 10, and keep doing this until at least one of the numbers you are multiplying is easy to deal with.

For example:

$8,000 \times 0.02$

$= 800 \times 0.2$

$= 80 \times 2$

$= 160.$

Or:

$0.2 \times 0.4 = 2 \times 0.04 = 0.08.$

DIVIDING BY LARGE NUMBERS

For division, the easiest method is to cancel out zeroes (i.e. keep dividing by 10) on both sides of the equation until you are just dividing by a single digit. So, for example:

$12,000 \div 40$ becomes $1,200 \div 4 = 300$

And:

$88,000 \div 300$ becomes $880 \div 3 = a\ bit\ less$
$than\ 300.$

When dividing by decimals, you can multiply both numbers by 10 until you are no longer dividing by a decimal. For example:

$0.006 \div 0.02$

$= 0.06 \div 0.2$

$= 0.6 \div 2$

$= 0.3$

TEST YOURSELF

(a) $1,000 \div 20$
(b) $6,300 \div 90$
(c) $160,000,000 \div 80$
(d) $2,200 \times 0.03$
(e) $0.05 \div 0.001$

Answers on page 192.

USING 'STANDARD FORM' FOR LARGE NUMBERS

Scientists often find themselves measuring vast or tiny quantities, and when calculating with these numbers they prefer to use what's known as standard form. This means expressing all numbers as a single digit followed by the appropriate power of 10. For example, 400 in standard form is 4×10^2.

The powers of 10 are then added (for multiplication) and subtracted (for division).

For example:

$$90,000 \times 40$$
$$= 9 \times 10^4 \times 4 \times 10^1$$
$$= 36 \times 10^5 \ (or \ 3,600,000).$$

and

$$700,000 \div 200$$
$$= 7 \times 10^5 \div 2 \times 10^2$$
$$= 3.5 \times 10^3$$

TEST YOURSELF

(a) What is 4×10^7 when written out in full?
(b) What is 1,270 written in standard form?
(c) What is 6 billion written in standard form?
(d) $(2 \times 10^8) \times (1.2 \times 10^3)$
(e) $(4 \times 10^7) \div (8 \times 10^2)$
(f) $(7 \times 10^4) \div (2 \times 10^{-3})$

Answers on page 192.

STAR WARS POWER

There's a 'standard form' joke that is told about Ronald Reagan's Strategic Defense Initiative (SDI) of the mid-1980s. I'd love to believe that it really happened.

The idea of the SDI, which was given the nickname 'Star Wars', was to develop laser weapons that would be capable of destroying enemy nuclear missiles at long range. The laser weapons would need a huge amount of energy, and millions of dollars were allocated towards researching the feasibility.

At one point during the research, the labs were asked to report back to the government.

'How much power will one of these weapons need?' asked one official.

'We're going to need 10^{12} watts, sir.'

'And how much can you deliver now?'

'About 10^6 watts, sir.'

'OK, good,' said the government official, 'so we're about halfway.'

In case you missed it, 'halfway' to 10^{12} would actually be 5×10^{11}, so the official was out by a factor of 500,000.

KNOW YOUR MEGAS FROM YOUR TERAS

The powers of 10 have been given official prefixes that you'll often encounter in discussions about energy and computer power in particular. Here's the background to their names.

		SI Prefix	Origin
10^3 (1,000)	Thousand	Kilo	From *chilioi*, meaning 'thousand', introduced by the French in 1799.
10^6 (1,000,000)	Million	Mega	From *megas*, meaning big or tall. First used as a prefix in late Victorian times.
10^9	Billion	Giga	From *gigas*, meaning giant. Officially adopted in 1960.
10^{12}	Trillion	Tera	From *teras*, meaning 'monster'. 'Tera' is similar to the Greek *tetra* (four) and, coincidentally, this is the fourth prefix.
10^{15}	Quadrillion	Peta	Adopted in the 1970s, this fifth prefix is a made-up word that is a nod to the Greek number *penta* but with a consonant left out, copying the pattern from *tera*.
10^{18}	Quintillion	Exa	Adopted at the same time as peta, this is *hexa* with the h removed.
10^{21} 10^{24}	Sextillion Septillion	Zetta, Yotta	These prefixes aren't in common use yet, but as computer power grows, you'll encounter them more often.

KEY FACTS

To be equipped to do back-of-envelope calculations, there are a few basic statistics that are valuable foundations from which you can build your estimates. Here are some important ones:

World population	Between 7 and 8 billion
UK population	Approaching 70 million
Distance London to Edinburgh (as the crow flies)	330 miles
Circumference of the equator	Around 24,000 miles
Walking speed of a commuter	3–4 mph (a bit below 2 metres/second)
Fastest that a human can run	A bit over 10 metres/second (~20 mph)
Size of a top Premier League football crowd	60,000 is typical
Cruising speed of a regular passenger jet	500–600 mph
Ceiling height of a typical apartment room	2.5 metres/8 feet
Fuel economy of a typical family saloon	40 miles per gallon
Weight of a litre of water	1 kilogram (exactly!)
Weight of a four-seater family car	A bit more than 1 tonne, less than 1.5 tonnes

TEST YOURSELF

Using the key facts above as a baseline, you can start estimating other things. Have a go at these:

(a) How far is it from London to Auckland in New Zealand?
(b) How far is it from London to New York?
(c) How many people live in Mexico City?
(d) How tall is a 20-storey office building?
(e) How long would it take a healthy adult to walk 10 miles?
(f) How many children attend primary school in the UK?
(g) How many people get married each year in the UK?
(h) What is the area of the Atlantic Ocean?

Answers on page 192.

ZEQUALS

If you have mastered all the other tips in this section, you are now well equipped to do a wide range of rapid estimations without needing a calculator.

There's one final tool to add: Zequals.

One of the secrets to doing quick back-of-envelope calculations is being able to make the calculations as simple as possible. There are many approaches to estimating, but Zequals is one of the most ruthless and is designed to minimise your need for a calculator. I named it Zequals, and because it has strict rules, I invented a symbol for it, too.

The idea behind Zequals is to simplify all the numbers you are dealing with before you do any calculations, by rounding them to one significant figure. In other words, you are rounding numbers to the nearest unit, or the nearest ten, hundred or thousand – always, and without exception.

The symbol for Zequals is $\approx$. Here are some examples of how Zequals operates. Notice how in each case, the number you start with is being rounded so that it only has one digit that is not zero:

$6.3 \approx 6$

$35 \approx 40$ (the Zequals convention is that if the second digit is 5, you round upwards)

23.4 ～ *20*

870 ～ *900*

1,547,812.3 ～ *2,000,000 (two million)*

Single-digit numbers and numbers with only one non-zero digit stay the same, because they already have one significant figure; so:

7 ～ *7*

0.08 ～ *0.08*

9,000 ～ *9,000*

Why Zequals? The Z stands for zeroes, because this method uses a lot of them. And the zig-zaggy symbol also looks a bit like a saw, which is appropriate, since this technique is a little like brutally sawing off the ends of numbers. And why is Zequals useful? Because it makes complicated calculations so manageable that you can do them in your head, and as we'll see, it usually takes you to an answer that is in the right ballpark.

Rounding numbers for doing estimations is nothing new, but if you are using Zequals, you have to stick to the rules at all times. And because this is rough-and-ready calculation, you should really 'zequalise' the answer too. So:

4 × 8 = 32, but then *32* ～ *30.* So *4 × 8* ～ *30*

A NEW MATHEMATICAL SYMBOL 〰

The symbol ≈ means 'is approximately equal to' and, unusually for maths symbols, there is no hard-and-fast rule for how to use it. For example: 7.3 ≈ 7.2. But it's also true that 7.3 ≈ 7.0 and that 7.3 ≈ 10.

The approximate value that you choose, be that 7.2, 7.0, 10 or something else, depends on your own judgement of what is most appropriate at the time.

Zequals, on the other hand, has very specific rules. Always and without exception, 7.3 〰 7, because it means 'round the number on the left to one significant figure'.

TEST YOURSELF

What are the following, according to Zequals?

(a) 83 〰
(b) 751 〰
(c) 0.46 〰
(d) 2,947 〰
(e) 1 〰
(f) 9,477,777 〰

Answers on page 195.

CALCULATING WITH ZEQUALS

Suppose somebody asks you how many hours there are in a year. Just roughly. There are 365 days in the year and 24 hours in a day, so that's 365 × 24. That's hard to work out in your head. But zequalise it and it becomes an easy calculation: 365 × 24 ≈ 400 × 20 = 8,000. Compare that with the exact answer of 8,760. It's only about 10% off, certainly in the right ballpark for most situations where you'd need to know the answer.

Using Zequals, long division, the bane of many school-children's lives, becomes a relative doddle.

What's 5,611 divided by 31?

It zequalises to 6,000 divided by 30, which is 200. Again, that's not far away from the exact answer (181). Zequals can be useful even when you are using a calculator because you need a more accurate answer. If your calculator is telling you the answer is 18.1, then your estimate using Zequals tells you that the calculator is wrong (most likely because you inadvertently pressed the wrong key at some point).

TEST YOURSELF

Work out the following using Zequals (and you might want to check how close your answer is to the precise answer, and whether your answer is too high or too low):

(a) $7.3 + 2.8$ ≈

(b) $332 - 142$ ≈

(c) 6.6×3.3 ≈

(d) 47×1.9 ≈

(e) $98 \div 5.3$ ≈

(f) $17.3 \div 4.1$ ≈

Answers on page 195.

THE INACCURACY OF ZEQUALS

I can't emphasise enough that Zequals is not designed to give you the exactly the right answer. In fact it can sometimes take you quite a long way from the right answer, particularly because of the Zequals rule that if the second digit is 5 you always round upwards.

Just how inaccurate can it be?

Take the example $35.1 + 85.2$. Rounding these to the nearest whole number would give you the answer 120, which is almost exactly right. But according to Zequals, this becomes $40 + 90 = 130$, which is nearly 10% too high. What's more, 130 ≈ 100, which is almost 20% too low.

In multiplication it can be a lot worse.

$$35 \times 65 = 2,275$$

But according to Zequals: $35 \times 65 \approx 40 \times 70 = 2,800$, and $2,800 \approx 3,000$. That's over 30% too high. Does this matter? It depends on what level of accuracy you are looking for.

TEST YOURSELF

(a) Multiplying together any two numbers between 1 and 100, what is the biggest over-estimate you can make if you use Zequals?
(b) And what is the biggest under-estimate?

Answers on page 195.

As you'll find from the Test Yourself box above, if you're unlucky with the numbers you are dealing with, Zequals can give an answer that is twice or half the right answer – and the more numbers you are putting into a calculation, the greater the scope is for deviating from the target.

At this point, an experienced estimator might want to use a different, more accurate approach. If both numbers are being rounded up to the nearest 10, a common method is to round down one of them to compensate. So, for example, when estimating 35×65, instead of calling it 40×70, it will be more accurate to round it to 30 × 70. And that is completely sensible if you are looking for a more accurate estimate (and don't want to use a calculator).

But let's not forget what the aim is here. What we are looking for is answers that are in the right ballpark. In many situations, 'the right ballpark' means 'the right order of magnitude'; in other words, is the decimal point in the right place? Zequals is all you need in these circumstances – and of course it has the great advantage that it reduces all calculations to being so simple that – with a little practice – you can do them quickly in your head.

And there is perhaps a more important point to make here, which might almost sound like a paradox.

The purpose of Zequals is to make every calculation so simple that almost anybody can do it. And yet knowing *when* it is appropriate to use Zequals, and knowing how to interpret the results, requires a degree of wisdom and a reasonable confidence with numbers. The better you become at arithmetic, the better you get at using Zequals.

WHO WANTS TO BE A MILLIONAIRE? (PART 1)

September 2001. It was a Celebrity Special edition of Who Wants to be a Millionaire? *Jonathan Ross and his wife Jane had made it through to the tenth question. They had £16,000, but had used up their lifelines.*

This was the question that would take them to a guaranteed £32,000: 'Which episode number did Coronation Street *reach on 11 March 2001?'*

(a) 1,000
(b) 5,000
(c) 10,000
(d) 15,000

The conversation went as follows:

Jonathan: *'So 50 weeks in a year. Twice a week. 100 a year.'*

Jane: *'There's not 50 weeks in a year.'*

Jonathan: *'It's 52 weeks in a year. But roughly. I'm rounding off so the audience can keep up. So it's about 104 a year. And about 40 years. So that's . . . lots. Should we go with (c) 10,000?'*

Jane: *'That's a ridiculously huge amount . . . and it can't be (d).'*

Jonathan: *'It's (b) or (c), I've talked myself out of (d). We're going to gamble £15,000 of someone else's money on . . . (c), 10,000.'*

Chris Tarrant: *'You had £16,000 . . . you've just lost £15,000.'*

Jonathan Ross began by doing everything right in his back-of-envelope estimation. His instinct of 'about 40 years' of Coronation Street was right. And he was sensible to round the weeks in a year to 50 for simplicity, making it 100 episodes per year, and 100 × 40 = 4,000 would have pointed the couple to the right answer, which was 5,000 (answer (b)). But shifting to the more accurate 52 weeks (making 104 episodes per year) was a distraction, and they never did do that final calculation. It's an example of where Zequals would have paid off.

3

EVERYDAY ESTIMATION

ESTIMATION AND MONEY

SHOPPING BILLS AND SPREADSHEETS

One of the most familiar uses for back-of-envelope maths is in adding up bills. If you're working to a budget and doing a big shop, it can be useful to mentally tot up how much you've put in the basket so far, so as to avoid getting a nasty surprise at the till. But there are also times when you will see a column of figures – in a bill, or in a spreadsheet[13] – look at the total and think: 'surely that can't be right'. In both cases, speedy ballpark estimates are useful.

One simple way to speed up the estimate of a shopping bill is to ignore the pence and just add the pounds. The result will be a figure that is an under-estimate – what's known as a *lower bound*. You can repeat the task, this time rounding odd pence *up* to the nearest pound, to give you the *upper bound*.

[13] Spreadsheets are a minefield for potential errors. One of the most common is when a formula is entered to 'sum' a column, but using the wrong range of cells, so that cells at the top or bottom are missed from the total.

The true figure will be somewhere in between the two, and a reasonable guess is to go for the middle figure.

Lower Bound	£32
Upper bound	£40
Estimate	£36

Quicker still, add up the pounds, and then add 50p for each item on the bill, so if the pounds add up to £32 and there are 10 items:

$$£32 + 10 \times 50p = £37.$$

What this doesn't allow for is the tendency for some products (books, for example) to have a disproportionate number of prices that end with 99p (e.g. £2.99 or £9.99) – or more commonly these days, 95p. This is a device used by retailers to trick us into thinking a product is cheaper than it is. The same items priced at just a penny higher, £3 and £10, feel much dearer, because we are more influenced by the leading digit than the later digits.

Rather than trying to allow for the peculiar distortions of supermarket prices, you can simplify things considerably just by rounding them to the nearest £1 or by using Zequals.

TEST YOURSELF

Bob has a spreadsheet that enables him to keep tabs on sales of widgets around the country. Here's the column for widget sales in Newcastle:

	SALES
	£ 190.10
	£ 120.46
	£ 8.22
	£ 396.63
	£ 130.50
	£ 41.55
TOTAL	**£ 697.36**

Do you trust the total at the bottom?

Answers on page 196.

HOW CAN THAT SHOP STILL
BE IN BUSINESS?

A few years ago, a new cooking utensil shop opened in my local high street in London. It was at the premium end of the street, where rents had been hiked up extortionately in recent years, and the gossip was that an annual rent of £25,000 was typical for a shop that size. It prompted me to wonder what their business model was.

If the rent is £25,000 per year, that's £25,000 ÷ 50 = £500 per week, or about £100 per shopping day. So just to pay the rent, the shop needed to make a profit of £100 per day. If the shop had a 100% mark-up on products, then it needed £200 per day in sales just to cover the rent (in other words, about seven of their upmarket saucepans). That was before the business rates, maintenance and insurance. It meant that before the owners could pay themselves and any other staff, they were probably having to put over £400 through the tills each day, on average. Yet frequently when I went past, there was nobody in the shop. 'How do they stay in business?' I wondered.

My question was answered not long afterwards, when the shop closed.

SAVING, BORROWING AND PERCENTAGES

Percentages are ubiquitous when dealing with money. I covered discounts and VAT on page 50, but percentages get more involved when you are dealing with loans and saving – and the numbers when working out what interest you will have to pay on a mortgage, or receive from an ISA, are probably going to dwarf those you encounter during shopping.

The maths, fortunately, is the same (5% interest on savings of £3,000 means you'll earn £150 each year). The complication comes with compounding: if your loan or savings run beyond a year, then you start paying, or earning, interest on interest.

If you have savings of £10,000 in an account that is paying 10% interest (happy days!), you will have £11,000 after one year. But at the end of the second year you won't have £12,000 because you are now earning interest on £11,000 rather than £10,000. After two years, your savings pot has therefore risen to £10,000 × 1.1 × 1.1 = £12,100. An extra £100. It doesn't sound much, but this margin becomes more significant the longer you save for, and the higher the interest rate is. (And on the other side, debts that are incurring compound interest can rise alarmingly.)

When interest rates are small (2.5%, say), there's a handy back-of-envelope rule for working out longer-term savings that's close to the right figure. If you are saving for four years at 2.5%, then the interest you've earned after four years is very close to 4 × 2.5% = 10%. (For comparison, the correct figure is 10.38%.) The smaller the interest rate, the better this approximation becomes.

This rule of small numbers allows you to be quite cavalier in calculations where you aren't required to give a precise answer, because it means you can just add and subtract percentages instead of multiplying and dividing them.

For example, if your savings go up by 3.3% in one year, 3.1% the next, and 2.7% the year after that, you're not far off the right number if you say that the total growth over three years will be 3.3 + 3.1 + 2.7 = 9.1% (and using Zequals, you can simplify it further: 3% + 3% + 3% = 9%).

This is fine over short periods. But over longer periods there's another handy rule of thumb: the Rule of 72.

DOUBLE YOUR MONEY – THE RULE OF 72

If your bank pays you compound interest at 4%, how long will it be before you have doubled your money?

This complex-sounding calculation can be answered with a deceptively simple rule. It's known as the Rule of 72.

Whatever the growth rate (be that 1.2%, 4%, 10% or even 30%), the time it will take for the quantity in question to double can be found by dividing it into 72.

With an interest rate of 4%, your money in the bank will double in $72 \div 4 = 13$ years.

How incredibly convenient (you might be thinking) that the number in this rule of thumb is 72, for 72 is a number that divides exactly by 2, 3, 4 and 6: numbers that will often be used as interest rates.

It turns out that, strictly speaking, this should be known as the rule of 69.3. That is the figure that emerges from doing the algebra behind exponential growth (described in more detail on page 181). But try dividing anything into 69.3 and you'll end up with a mess. Whoever first worked out this rule quickly spotted that by nudging it up to 72 instead, there was a chance people would be able to work out the numbers on the back of an envelope – or even in their heads. So the Rule of 72 it is.[14]

Knowing how long it will take for numbers to double is handy, but there may be times when you want to know a different target. What about trebling your money, or increasing it tenfold? It turns out there is a rule of thumb for any target of growth that you choose. In each case, it uses a convenient number that is quite close to the accurate one.

Here's a table:

How many years before a quantity . . .	Convenient number to divide into	Example: How long it takes if growth rate = 4%
Doubles	72	$72 \div 4 = 18$ years
Increases threefold	120	$120 \div 4 = 30$ years
Increases tenfold	240	$240 \div 4 = 60$ years

[14] The Rule of 72 applies to other things that grow, such as populations. If the world population is growing at 1.2% per year, it will be $72 \div 1.2 = 60$ years before there are twice as many people on the planet as there are now.

CONVERTING CURRENCY

Any international trip or purchase is going to involve a conversion, and unlike units of measurement, the conversion rates between sterling, euros and US dollars are changing all the time. During the current century, £100 could have bought you anything between $120 and over $200, which is a massive variation.

With many currency conversions, you – or the person you are dealing with – are probably going to be concerned with calculating figures that are correct to the last penny, or cent, and you'll almost certainly use a calculator. But imagine that you are at the airport and need some US dollars. The exchange rate is at $US1.40 to the pound, you ask the travel-exchange desk for $1,000, and the cashier charges you £793.40. Happy? Well, £793.40 is about £800, and a mental check tells you that 1,000 ÷ 800 = 1.25, which is a long way from the supposed $1.40 rate. So they're either charging a huge commission, or the cashier has mistyped a number. Do you still want to go ahead with this transaction?

Fortunately for the British, the pound is more valuable than most of the rest of the world's units of currency,[15] meaning that £1 will usually buy you more than one dollar, euro, Swiss franc, and far more yuan or rupees. Therefore, converting sterling to other currencies generally means multiplying by a number between 1 and 2.

You probably have your own ideas of mental short cuts

---

[15] This is only because of the historical decision of what level to set the pound at. It does not reflect the relative wealth of the countries!

you might use to do a rough conversion, but depending on the exchange rate, you might round to the nearest convenient ratio. For example:

Exchange rate of other currency	Close to ...	Estimation Short cut
1.09	1.1	Add 10%
1.35	1.33	Add one-third
1.52	1.5	Add a half
1.72	1.75	Add three-quarters
1.81	1.8	Double, then take off 10%
2.1	2	Just double it!

This is fine if you are converting from (say) sterling to US dollars. But if you are doing the reverse, it will mean multiplying by a number less than one, or dividing, and most people find both of these harder to do mentally. A confident arithmetician might be happy dividing by (say) 1.4, but the back-of-envelope approach will be to divide by 2 and add about 50% to the answer. And that's probably going to be good enough for when you need it. 'So that hotel is going to cost us $500 for one night – let's see, halve that = £250, plus £125 ... more than £350: ouch! – way beyond our budget.'

ESTIMATING SIZE

ESTIMATING DISTANCES

The most straightforward way to estimate distances is by comparing the distance you are trying to figure out to one that you already know. If you know that the distance from Sydney to Melbourne is 500 miles, then the distance from Sydney to Canberra – a city that for historical, political reasons needed to be roughly in the middle of the two cities – is going to be about 250 miles. And if you don't know the distance from Sydney to Melbourne, you might be able to estimate it using some other information you have – for example, that it takes about an hour to fly from one to the other. Since most planes fly at about 500 mph, in one hour the plane will cover 500 miles. There is, of course, plenty of 'ish' involved here.

The same principle applies to shorter distances, of course. If you know that your own height is 1.5 metres (say), then you can estimate the height of the room you are in by, for example, picturing whether you'd reach the ceiling if you stood on your own shoulders. (I'm a firm believer in picturing things like this.)

All of this is fairly routine and familiar. But there are three rather more quirky (and I think charming) ways of estimating distance and height.

1. The Dropped-Stone Method

When I was a child, we'd often take a trip to Beeston Castle in Cheshire. One of my favourite features there was the well, and we'd delight in dropping a pebble into the well and counting the seconds before we heard it hit the bottom. (With decades-worth of children doing the same thing since, I fear the well is no longer quite as deep as it was.)

You can estimate the depth of the well with a bit of Newtonian maths, which says that the distance travelled by an object falling under gravity is given by:

$$Distance = \tfrac{1}{2}\, a.t^2$$

where a = the acceleration due to gravity (about 10 metres per second per second) multiplied by the square of the time. If the pebble drops for three seconds, the depth of the well is therefore roughly:

$$\tfrac{1}{2} \times 10 \times 3^2 = 45 \text{ metres.}$$

There are two things being ignored here that mean the measured time will be an over-estimate. The first is that, because of air resistance, the pebble will eventually stop accelerating, and will therefore take longer to hit the bottom than it would have done in a vacuum. The second is that when the pebble hits the bottom of the well, it takes time for the sound to travel back up to your ears. Both of these effects are quite small, however. So, for a decent estimate of the height, square the time and multiply by 5.

2. The Finger Method

Let's say you are standing on the beach and can see a yacht on the horizon. You wonder how far away the yacht is.

Here's one way to find out. Stretch out your arm in front of you, close one eye and hold up a finger so that it covers the yacht.

Now open that eye and close the other one. Your finger will appear to jump to the side, away from the yacht (a phenomenon known as 'parallax').

The distance to the yacht is roughly:

10 × the distance that the yacht jumped.

You will need to use your judgement to estimate by how many yacht-lengths the yacht has jumped to the side. If you estimate that the yacht jumped by about 15 times its own length, then the yacht is roughly:

10 × 15 yacht-lengths
= 150 yacht-lengths away from you.

There is of course one other thing you need to estimate: the length of the yacht itself. You'll need to use your rudimentary knowledge of yachts to decide if this looks like a 5-, 10- or 20-metre yacht. If you reckon it's a 10-metre vessel, then your estimate is that it is about $10 \times 150 = 1,500$ metres away.

As I say, it's quirky. But I bet you now want to try it with that pylon you can see out of the window . . .

3. The Crisp-Packet Method

You are in the park and you are curious to know how tall a particularly fine poplar tree is. You can make a good estimate using an empty crisp packet.

Fold over the corner of the packet so that the top of the packet lines up with the side, and make a crease on the diagonal where the fold is. This diagonal is now at 45 degrees.

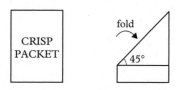

Hold the crisp packet next to your eye, and look along the diagonal as if it's a telescope, keeping the bottom of the crisp packet horizontal.

Walk towards the tree until the diagonal of the crisp packet lines up with the top of the tree.

Now, taking large strides of about one metre, count how many strides it is to the base of the tree.

The height of the tree is approximately:

the number of strides + your height.

How does this work? By lining up the folded crisp packet with the top of the tree, you have formed an isosceles triangle: the distance from you to the tree is the same as the height of the tree above your eye-line.

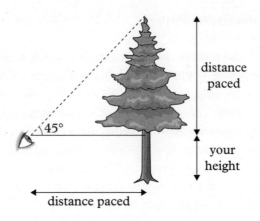

CIRCLES AND PI

There are two formulae related to circles that every child is required to learn.

For a circle of radius R:

$$Circumference = 2\pi R.$$
$$Area = \pi R^2.$$

But what value is π? It depends who you ask.

A mathematician will tell you that pi is the ratio of the circumference of a circle to its diameter, a transcendental number that begins 3.14159 . . . and continues for ever.

An engineer (so the joke goes) will tell you 'π is about 3, but let's call it 10 just to be on the safe side'.

Whichever of these views you sympathise with, when it

comes to most real-world problem solving, knowing that $\pi \approx 3$ is good enough.[16]

But when on earth might you need to use it at all?

In the 2004 Olympics in Athens, British athlete Kelly Holmes won gold in the 800 metres. Five days later, she had made it to the final of the 1500 metres, and she was aiming to become the first British athlete to win gold medals in both distances.

Kelly Holmes was a tactical runner, and was prepared to run at a pace that was comfortable for her, even if it meant she spent some of the race at the back of the field. As the final lap began, she was positioned in eighth place. Now she had to get to the front. The problem was that, in overtaking, she would need to be in the second lane and run outside the athletes in front of her. On the straights this would make no difference, but both ends of the track are semicircles, and Holmes therefore had to run round a circle with larger circumference than her competitors. In other words, to win gold, she had to run further than 1,500 metres.

But how much further?

At first glance it would appear that we don't have enough information. How long is an Olympic race track? What's the radius of the bends? How wide are the bends? But it turns out that only one of these items of data is important.

---

[16] A commonly used approximation for pi is $22 \div 7$, which is remarkably accurate (it's within 0.04% of the correct value of pi!). But if you want to learn pi to more decimal places, there are several mnemonics, of which my favourites are: and 'How I wish I could calculate pi' and 'May I have a large container of coffee.' In each case, count the number of letters in each word to give the digits of pi to several decimal places.

Take a look at the sketch of a track. Let's call the length of the straights L, and the radius of the inside lane at the end R. Remembering that the circumference of a circle is $2\pi R$, the length of a lap is twice the length of the straights plus the circumference of the circle, or $2L + 2\pi R$. But Kelly Holmes had to run around a circle whose radius was larger – by the width of one lane.

How wide is a lane on an athletics track? Just picture it in your mind. Thirty centimetres? No, much wider than that. A couple of metres (the equivalent of an athlete lying across it)? No, less than that. A metre sounds about right.[17] So let's say the radius of Kelly's circle was R + 1.

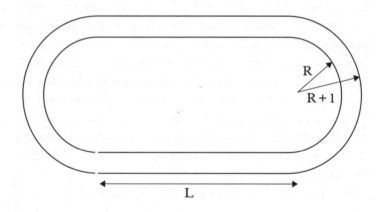

---

[17] The width of a lane is actually 1.22 metres. Like most dimensions in sport, it was originally specified in a nice round number of imperial units: 1.22 metres = 4 feet.

We can now work out the length of Kelly's lap:

$$2\pi(R + 1) + 2L$$
$$= 2\pi R + 2\pi + 2L.$$

Subtract from it the length of the inside lap and Kelly's 'extra' distance is:

$$= 2\pi R + 2\pi + 2L - 2\pi R + 2L.$$

$2\pi R$ and $2L$ cancel out to leave us with 2π . . ., which is 2×3.14 . . . let's call it 6 metres (since everything is an approximation here).

Six metres – that's a lot. It's the difference between a gold medal and being an also-ran.

What's interesting is that Kelly Holmes must have built this into her tactical calculations for the race: she knew she'd have to run further, but felt it was a price worth paying for enabling her to run at her own pace. And it worked: she won the race with a couple of metres to spare.

And that is how Kelly Holmes became *Dame* Kelly Holmes.

AREAS AND SQUARE ROOTS

We are often presented with numbers that are in square units, particularly area. A description of an apartment might say that it is '1,200 square feet', while a forest fire might be said to be covering '100 square miles'. Picturing square anything is difficult – we tend to find it easier to think in lengths. A hundred square miles is the equivalent of a square with sides that are 10 miles long, and to find the length of the side of that square, we need to be able to work out the square root of the area.

Here's a real example. In the winter of 2013–14, the South West of England experienced one of its wettest months ever. As a result, an area of low-lying land known as the Somerset Levels was flooded, and the flood waters remained for several weeks. At its peak in January 2014, it was reported that 69 square kilometres had been flooded.

Let's picture how big this area of flooding would be if it were a square.

If the area of a square is 69 km^2 then the length of each side is

$$\sqrt{69}$$

which is a number between 8 and 9 (and nearer to 8). So, the area that was flooded was roughly the same as a square that was 8 km × 8 km, or 5 miles × 5 miles. Now that is something that I *can* just about picture.

The Somerset Floods story was an example of where it could be handy to be able to work out the square root of a number. Working this out exactly can be messy, but there's a neat method for making a good estimate.

Suppose you want to work out the square root of 170,423.

Starting from the right-hand side of the number (the units column) and working leftwards, break the number up into pairs of digits, like this:

17 04 23.

Start with the first pair of digits (17) and estimate the square root of that number. Since the square root of 16 is 4, the square root of 17 is going to be 4-and-a-bit. If we're using Zequals, we just call it 4, but if we want a little more accuracy we can call it 4.1.

Now count how many other pairs of digits there are, and for each pair, multiply the square root of the first number by 10. In this case, there are two other pairs, so we multiply $4.1 \times 10 \times 10 = 410$. So the square root of 170,423 is about 410.

Try another: the square root of 4,138,947.

Split the number into pairs, starting at the right:

4 13 89 47

(notice this time that the opening 'pair' is just a single digit: 4).

The square root is therefore, roughly:

$2 \times 10 \times 10 \times 10 = 2,000$.

TEST YOURSELF

Can you estimate in your head the square roots of the following numbers? If you end up within 5% of the exact answer, give yourself a point. Within 1%, give yourself a hefty pat on the back.

(a) 26
(b) 6,872
(c) 473.86 (hint – ignore the digits after the decimal point!)
(d) The floor area of a flat is advertised as being '910 square feet'. How big would that be if it were a single square room?
(e) According to Wikipedia, the Caspian Sea is 371,000 km². If it were a square with the same area, would it fit inside the borders of France?

Answers on page 196.

WHO WANTS TO BE A MILLIONAIRE? (PART 2)

The year was 2008,[18] and this was an episode of Who Wants to be a Millionaire? featuring couples. One couple – let's call them the Smiths – had made it to £64,000.

The question that would take them to £125,000 was this: 'Which ocean has an area of 4.7 million square miles?'

(a) Arctic
(b) Atlantic
(c) Indian
(d) Pacific

The Smiths didn't know the answer, so decided to use their final life-line, which was 'Ask the audience'.

About half the audience voted for the Pacific, but the couple had hoped for a more definitive vote – 80% or more – so they decided to play safe and take the money.

The reason I know about this story is that my friend John Haigh, a lecturer at Sussex University and keen back-of-enveloper and Zequaliser, told me the next day that this question had cropped up, and that he had worked out the answer in his head. As his starting point, he made an estimate of the size of the ocean with which he was most familiar: the Atlantic. Can you figure out which was the correct answer? (See page 182)

---

[18] Details of this episode seem to have been lost in the archives; this is our joint best attempt at recalling the details of what we saw.

METRIC AND IMPERIAL CONVERSIONS

WHO NEEDS IMPERIAL?

Like it or not, we will all continue to need to convert from metric to imperial and vice versa for some time yet. Why? There are two reasons:

(1) **The USA:** The world's biggest economy and most influential culture still stubbornly talks and works in feet, yards, pounds and gallons. This rubs off on the rest of the world because there will be references to imperial units not only in engineering technical specifications but also, for example, in popular songs, movies and American cookery books.

(2) **Old habits die hard:** Commonwealth countries that shifted from imperial to metric still have a legacy of imperial measurements in their language and their thinking. For example, in New Zealand, which went fully metric in 1976, the number of kilometres that a car has travelled is still referred to as its 'mileage'. But that's nothing compared with the UK, which is divided down the middle as a nation on which system it uses. Even individuals have split personalities when it comes to measurement. I have met numerous adults who, for example, know their height in feet but their weight in kilos. And this is not just an age thing. I've done surveys of hundreds of 15-year-olds across the UK, and the results are consistent:

- About 75% of 15-year-olds estimate other people's height in feet and inches.

- About 30% of them estimate other people's weight in stones and pounds.

This widespread use of imperial units is despite the fact teenagers never encounter these units in their school exams, and despite the fact that almost none of them know that there are 14 pounds in a stone! So, in the UK, whichever units of measurement you use, you are going to encounter people who prefer the other units.

THE *MARS ORBITER* FIASCO

In December 1998, NASA launched a space probe called the Mars Climate Orbiter. *Its mission: to study the Martian atmosphere. Several months later, as the orbiter approached the planet, it fired the thrusters that were designed to put it into a stable orbit. But to the horror of the NASA team who were monitoring progress, the rocket thrusters were much too strong and the probe hurtled into the planet and was destroyed.*

A NASA review board later discovered that the software designed by the Jet Propulsion Laboratory at NASA had used the metric system in its calculations, but the engineers at Lockheed Martin Astronautics who built the spacecraft had based their calculations on traditional inches and feet (in the report, this was referred to as the 'English system', as if this were somehow not the USA's fault). Instead of applying pound-force, the rockets applied Newton-force, about four times bigger. The cost of this simple error was $125 million of lost space probe.

MILES AND KILOMETRES

We took a huge step forward when we switched from imperial units to metric. Calculations in feet, pounds and gallons were simplified overnight when everything could be worked out in decimal.

In the UK, metrication really kicked in during the early 1970s, when we joined the EEC, in which metric units were already standard. The school curriculum went metric at the same time, which means that maths education of anyone under the age of 50 focused exclusively on metric units. With one, glaring exception: the mile.

Road signs are exclusively in miles, and, by default, every car speedometer is in miles per hour. We therefore have the curious situation where the vast majority of teenagers will quote long distances in miles, but short distances in metres; and they will quote faster speeds in miles per hour, but slower, more human speeds in metres per second. Confusing? Well no, not really. So long as they aren't having to switch between the two, most people are comfortable enough working in whichever imperial or metric unit they are used to in the context they are working in.

The problems start when there is a need to convert from metric to imperial. I asked a large group of 15-year-olds to estimate the distance from London to New York. Their answers varied considerably, but most of them were in the not-completely-outrageous range between 1,000 and 10,000 miles (the correct figure is around 3,500 miles as the crow – or at least the Boeing 787 Dreamliner – flies).

The problems arose when they were asked to then quote

that number in kilometres. There was relatively low awareness of what the relationship is between a mile and a kilometre, other than that they are both 'quite long'. In many cases, the mileage figure was just multiplied by 10 to get a figure in 'kilometres'. Perhaps this mistake arises because the shorthand for both metres and miles is 'm'. There's a vague awareness that an 'm' is related to a 'km', so why not multiply by 1,000 . . . no wait, that sounds too high . . . why not multiply by 10 instead?

The actual ratio of kilometres to miles is 1.6. (More precisely, it is 1.609 . . .)

How To Remember Your Conversions

These three mnemonics appeared on the back of Kellogg's cornflakes packets in the 1970s, and those of a certain generation have never forgotten them:

A litre of water's
a pint and three-quarters

Two and a quarter pounds of jam
weigh about a kilogram

A metre measures three foot three,
it's longer than a yard, you see.

BACK-OF-ENVELOPE CONVERSIONS

If it's too much of a faff to use the accurate ratios above, you can use a Zequals-style approach to give you conversions that will suffice in most situations. And conveniently, all of the common rough-and-ready conversions only require doubling or halving.

	Accurate conversion	Rough conversion	Example
Litres to pints	× 7/4 (or 1.75)	Double	10 litres ~ 20 pints
Litres to (UK) gallons	× 7/32	Quarter	20 litres ~ 5 gallons
Kilometres to miles	× 5/8	Halve	200 km ~ 100 miles
Metres per second to miles per hour	× 2¼	Double	10 m/s ~ 20 mph
Centimetres to inches	× 2/5	Halve	6 cm ~ 3 inches
Metres to yards	× 13/12 (add 1/12)	Equal	70 metres ~ 70 yards
Kilograms to pounds	× 2¼	Double	10 kg ~ 20 pounds
Celsius to Fahrenheit	× 9/5 and add 32	Double and add 30	20 °C ~ 70 °F

There are of course other more obscure imperial measurements that you might encounter, such as acres (land), furlongs (in horse racing) and fluid ounces (cooking), but these rarely crop up in everyday encounters and you're unlikely to have to deal with converting them on the hoof.

TEST YOURSELF

Do rough conversions of the following in your head:

(a) 70 miles in kilometres.
(b) 40 kilograms in pounds.
(c) 150 metres in yards.
(d) 100 kilometres in miles.
(e) 25 °C in Fahrenheit.
(f) 10 stones in kilograms (one stone is 14 pounds).

Answers on page 197.

A QUIRKY METHOD FOR MILE–KILOMETRE CONVERSION

About 800 years ago, a mathematician called Leonardo of Pisa (who was nicknamed Fibonacci) wrote about a curious sequence of numbers. Starting with 0 and 1, the sequence goes as follows:

$$0 \ 1 \ 1 \ 2 \ 3 \ 5 \ 8 \ 13 \ 21 \ 34 \ 55$$

Each number in the sequence is obtained by adding the previous two terms. So, after 55, the next term will be 34 + 55 = 89.

Now here is the remarkable thing. From the number 3 onwards, if you take any two consecutive terms in the Fibonacci sequence, their ratio is very close to 1.6. For example, 13 ÷ 8 = 1.625, and 34 ÷ 21 = 1.619. This isn't just a fluke; it turns out that as you go further along the sequence, the ratio of successive terms in the Fibonacci sequence gets closer and closer to a number known as the 'Golden Ratio', which is roughly 1.618.

The coincidence is that the Golden Ratio is very close to 1.609, which is the ratio of miles to kilometres. So if you want to convert 13 miles to kilometres, then, just by glancing at the Fibonacci sequence, you can estimate that the answer is going to be about 21 km, and you'll be correct to within 1%.

It works in reverse, too. Travelling around Europe, you spot that your destination is 34 kilometres. 'That's 21 miles,' you can state, with remarkable accuracy.

ESTIMATION AND STATISTICS

AVERAGES AND UNCERTAINTY

The word 'average' is used in everyday speak to mean 'typical' or 'somebody in the middle'. In many situations it's fine to use this general word, but it's worth being reminded that there are three different averages that are commonly used.

The mean is the most commonly used average. It's found by adding up all the values or measurements, and dividing by the number of items you are measuring. The mean is what's used when referring to average adult height, batting averages in cricket and also people's income.

The median is the middle value, if you were to line up all the data from smallest to largest.

The mode is the data value that crops up the most often. For example, the 'modal' shoe size for an adult woman in the UK is 6.

We've seen earlier that most statistics have an element of uncertainty, so that statistics that you are presented with might be an over- or under-statement of the true figure.

The cause of this 'error' will be one of two things: either the method you use to measure the statistic isn't reliable (a weighing scale that gives different readings each time, for example), or the thing that you are measuring tends to vary (for example, if you are looking to find the height of a typical person).

Either way, the 'true' figure is going to be somewhere on a spread of possible values. Most often this spread (more formally known as a distribution) will look something like this:

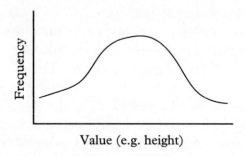

Value (e.g. height)

This shape is known as the Normal distribution (so called for the banal reason that it's not abnormal), though it's often called a bell curve (because it is shaped like a bell). Points in the central, higher region of the curve represent values or readings that arise most frequently, while the lower regions left and right are the more extreme and less frequent values. The heights of children in a class, the time it takes for daffodils to bloom and many other everyday phenomena follow this sort of pattern. It's handy that this spread is symmetrical, because it means the average (mean) value is right in the middle. In distributions like this, it's just as likely that a statistic or measurement will be higher than the quoted figure as lower, so it's fine to call the highest point the 'average'.

However, not all statistics follow this pattern. For example, if you were monitoring the amount of time that people spend in a toilet cubicle at a rock concert,[19] the distribution would look something like this:

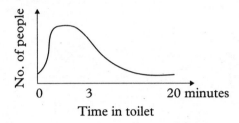

Most people might spend three or four minutes, but a few take 10 minutes or more, and one or two will exceed 20 minutes. This distribution is known as a 'log normal' (cue lavatorial jokes). Looking at this graph, you'd say that the typical time spent is two to three minutes, but the 'mean' (the total time spent divided by the number of people) is going to be higher than that because of the few extremes.

Adult income, similarly, has a skewed distribution, with many people clustered in the range £20,000–£30,000, but with a long tail to the right that includes a few on multi-million salaries. So, while the middle (median) income is around £25,000, the 'mean' income will be to the right of the peak. The majority of people earn less than the average (mean) income – which can make the choice of which average to quote very political.

[19] I know this because my friend Aoife Hunt had to do it, when researching crowd logistics.

There are two other distributions that are also quite common. If you take a random sample of tweets on Twitter, the number of 'likes' for individual tweets will look like this:

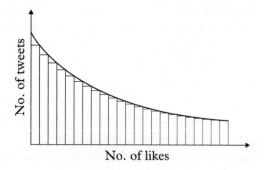

The most common number of likes is zero, followed by 1, then 2, each becoming less likely. This is known as an exponential distribution.

Finally, if you were to pick a more economically developed country and gather together everyone between the age of 20 and 50, the distribution of their ages will typically look like this:

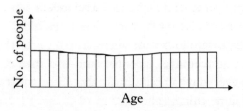

Yes, there will be bumps from year to year, and there might be a slight trend going up or down, but broadly speaking it's a flat line. If you pick somebody at random from this group, the chance of the person you choose being 23 will be about the same as the chance of them being 33.

You'll get a similar pattern with, for example, the time you expect to wait for a London Underground train if you turn up at a station at random. Trains might be spaced every couple of minutes, and you are as likely to arrive on the platform at the beginning of the two-minute gap (just after a train left) as you are to arrive at the end of the gap (as a train is pulling in). And, of course, you could arrive anywhere in between.

Knowing which of these distributions a particular statistic belongs to can be helpful in making estimates.

WORKING OUT PROBABILITY

'Probability' is the formal way of saying 'the chance of a something happening'. Probabilities range from absolutely certain (i.e. it will happen 100% of the time, such as the sun rising tomorrow) and impossible (0% of the time), and can be anything in between. The chance of flipping a head on a fair coin, for example, is 50%, while the chance of picking a Heart from a regular pack of cards is 25%, since a pack is divided equally into four suits.

Although percentages are the most common way of expressing everyday probabilities, there are other ways of saying the same thing. The chance of picking a Heart can be represented as:

- A fraction (¼, or '1 in 4').
- As a decimal (0.25).

- As bookmakers' odds (selecting a Heart is 3:1 against, so that for every time you pick a Heart, you can expect to pick a card that isn't a Heart three times).

But some probabilities can't just be stated by simple inspection. Sometimes, to find (or estimate) a probability, you need to take a different approach. For example:

- To estimate the probability that a car will be yellow, you could do a *survey*. Count the next 100 cars that you pass on the open road. If one of them is yellow, that suggests the chance of a car picked at random being yellow is about 1 in 100. The bigger the sample, the more accurate your estimate of the probability will be.
- To estimate the chance that it will be warmer than 25 °C in Paris next 1 July, you can use *history*. If it has been warmer than 25 degrees on 1 July in seven of the last 10 summers, in the absence of a weather forecast it's a reasonable guess that there's a $7/10$ (70%) chance that it will exceed that temperature this year too.
- Sometimes we just have an instinct for the probability of something happening based on a *hunch*. When answering the quiz question: 'Who was older when they became US President: George W. Bush or Bill Clinton?' you will have a gut feeling of what the answer is, to which you could put a value: 'I reckon it's 90% Bill Clinton' or 'a bit more than 50:50 it's Bush'. If you have absolutely no idea, then the chance you will get the right answer out of a choice of two is exactly 50%.

The more vague the probability that you are using in your calculations, the less reliable your estimates based on it will be. So, while you can work out precisely the odds on you getting a 'flush' of five cards in poker, you can only estimate very approximately the chance that your team will win a pub quiz.

BACK-OF-ENVELOPE SURVEYS

Many of the statistics that are fed to us are the result of surveys. If we're told that 7% of the public plan to vote for the Green Party at the next election, or that 65% of children spend over three hours a day at home in front of screens, those figures aren't based on a census of the whole population, but on a sample of perhaps 1,000 people, chosen carefully to be 'representative' of the population by age, gender, social background and so on. Thanks to the large sample, the pollsters and market researchers can give us a figure that they know is quite reliable.

But surveys don't have to be restricted to the professionals. You can do your own back-of-envelope versions. The experts would pull their hair out if you conducted a sample based on, say, 10 people, but even a tiny sample can begin to give you a sense of the bigger picture.

Several years ago, I was at a committee meeting where concern was expressed at the falling membership of a national body that I was working with. There was a proposal to hire some consultants to help draft a survey that could be sent out to all of the several thousand members, canvassing their views.

I had an alternative suggestion. We needed information quickly. Why didn't the 10 of us around the table each take a couple of copies of a short questionnaire, and the next time we met a potential member over the coming week or two, ask them to fill it in. I reckoned that even a handful of responses would give us a huge pointer to what the big issues were. I was voted down, but I decided to do my guerrilla research anyway. Out of five people that I talked to, two said that the reason they were not members was that they had no time to enjoy the benefits, and three said they had once been members but now met their needs using free alternatives that were available via social media.

With that tiny sample, it would be bogus to claim with confidence that '40% of the target group no longer have time to enjoy the benefits of membership' or that '60% now get their resources elsewhere'. But, in truth, we didn't need to know the results to the nearest 10%. If the result of the mini survey had been 90% and 30%, we would still have come to the same conclusion, which was that new threats had emerged, from time pressure and from social media, and that these had to be addressed. (That national body never did get round to its big survey – but it did make some positive changes, including engaging in social media.)

Back-of-envelope surveys are nothing special. We all do them, all the time. We ask a few friends which plumber they'd recommend, or what the going rate is for the tooth fairy when a child loses a tooth. A survey of three people will produce a result with a huge margin of error. And yet, in the end, we can still glean valuable insights, and make better decisions, if our survey tells us that 67% of the public (OK, in truth it's two out of three of our friends) say that in their house the tooth fairy pays £1.

Of course, you should do big, statistically rigorous and

representative surveys when you can, and when accuracy is important. But when you don't have the time or the money to do it, don't rule out the value of un-rigorous, biased, back-of-the-envelope alternatives – as long as you remember not to set too much store by the numbers in the results.

HOW LONG WILL WE BE IN THIS QUEUE?

It's October half-term. I'm at Legoland theme park. Again. The kids are desperate to go on the 'Pirate Falls' ride. My heart sinks when I see the sign saying the wait is currently one hour. That will be one hour shuffling slowly along the queue, with the prospect of a five-minute boat ride and a soaking at the far end.

Luckily, the queue gives good views of people setting off on the ride, so this is a chance to check out if Legoland's prediction of one hour is right. If it's really going to take that long, we'll go and find something else to do.

We decide to do a survey. We watch the boats setting off at the start of the ride, and over a period of five minutes we count how many people have gone past. Some boats have four people in them (Hooray, that will deplete the queue!), a couple have none (Boo! What a waste of a good boat).

Over five minutes we count 36 people, so we work out that the average throughput is about seven people per minute. Then we estimate the length of the queue – about 150 people.

One hundred and fifty people at seven people per minute – 150 divided by 7: that's about 20. Clearly Legoland's prediction of one hour is massively wrong; it will be more like 20 minutes. This cheers me up no end, though it turns out that there's a factor I hadn't allowed for: Q-bots, which allow premium customers to go straight to the front of the queue through a separate entrance. It turns out that Q-Bots slow the queue down by about 25%, so it's nearer 25 minutes before we get to the front of the queue.

Still, back-of-envelope thinking ensured we were better informed than those who trusted the waiting-time sign, and it also helped occupy us for a few minutes. I was happy. Until we got to the end of the ride, when we left the boat looking like drowned rats.

THE CHANCE OF MULTIPLE EVENTS

You can work out the probability of two or more *independent* events happening by multiplying their probabilities together.

It's often convenient to work out the probability of more than one event happening by using fractions – indeed, this is perhaps the most important application of the methods that you learned at school for adding and multiplying fractions. For example, the chance of getting two sixes when rolling two dice is worked out as:

$$\tfrac{1}{6} \times \tfrac{1}{6} = \tfrac{1}{36}.$$

That's a lot easier than working out 16.7% × 16.7%.

The events are independent if one of the events is in no way influenced by the other: rolling a dice and flipping a coin are independent events, but living in Wales and having the surname Jones are not. One in 20 people in the UK live in Wales, and around 1 in 100 in the UK have the surname

Jones – but the chance that the next winner of the UK Lottery is a person who lives in Wales *and* has the surname Jones is not $\frac{1}{20} \times \frac{1}{100}$ (= 1 in 2,000). The proportion of Welsh residents called Jones is around 1 in 17,[20] so the chance of a Lottery winner being Welsh and called Jones is going to be roughly

$$\frac{1}{20} \times \frac{1}{17} = 1 \ in \ 340.$$

Knowing whether or not two events are independent is partly common sense and partly experience, but for back-of-envelope purposes, as a starting point you can generally treat two events that don't have an obvious connection as being independent.

For example, suppose you're running a bit late, and need to catch a bus to the station where you want to catch a train. Suppose about four-fifths (80%) of trains leave your local station on time, and your hunch tells you that the chance you'll have to wait more than five minutes for a bus (and hence be late for the train) is about one in two (50%). Now, the chance of the bus being late and the train being late aren't entirely independent: bad weather would affect both of them, for example. But that connection probably isn't significant. The chance that you'll have to wait more than five minutes *and* that the train will be on time is therefore going to be about $\frac{4}{5} \times \frac{1}{2}$ which is $\frac{2}{5}$, or 40%.

---

[20] In 2008, 5.7% of the Welsh were called Jones according to the journal *Significance*.

SPOTTING TRENDS

Much of our modern life is underpinned by statistics. They dominate our news, and we use them to form opinions, make judgements and, most importantly, make decisions. It's a statistician's job to wade through data and to spot important patterns and connections. And in an era where 'big data' is used to help advertisers, political parties and other institutions to understand our behaviour in frightening detail so that they can understand and perhaps influence us, it should come as little surprise that the best statisticians can earn huge salaries.

The maths involved in statistics can get quite advanced. If you've got a set of data (such as the fictitious points in the graph below) and are looking for the straight line that most tightly passes through it (the so-called 'best fit'), there are sophisticated mathematical techniques you can use to find it.[21] But in many cases, the human eye is good enough. A bit like in a spot-the-ball competition, I've used my judgement, and a hunch, to draw a straight line through the points below. It suggests there's a slight trend upwards. Your line might be different, but it's unlikely to be *that* different.

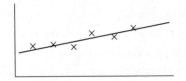

---

[21] One such is known as 'least-squares' regression.

Often, if you're looking to forecast what's going to happen in the near future, a straight-line extrapolation from the past is a good starting point. Here, for example, is a chart showing the proportion of all grocery spending that was done online from 2013 (2.1%) to 2017 (4.8%). Care to guess what happened in 2018?[22]

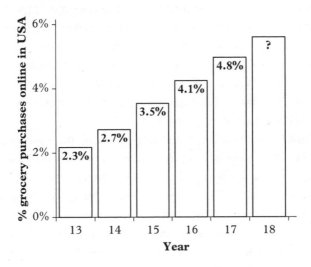

The online market was clearly growing each year, though not by a fixed amount. The annual growth was just 0.4% in 2014, and for the next three years it was in the range 0.6% to 0.8%. A guess of 0.7% growth seems sensible for 2017–18. Of course it might be as high as 1% or as low as 0.3%, or it might do something cataclysmic, we can't be sure, but like steering an oil tanker, statistics that are heading steadily in one direction can take a lot of effort to shift to a different

[22] Source of data: Global/Data analysis.

path. So a prediction of 5.5% (ish) for 2018 is a relatively safe bet – and as it happens, 5.5% is exactly what the growth was that year. But I've been a bit lucky here – informed guesses aren't always as accurate as this.

The further into the future you are trying to forecast, the more risky it is to extrapolate a trend from the data you have. And be careful: while the grocery purchasing habits of the USA are clearly revealing a changing pattern of consumer behaviour, it's possible for an 'upward trend' to happen purely at random. If I repeatedly toss 10 coins and get four heads the first time, five heads the second time and six heads the third, it might look like an upward trend, but the stats say that the most likely outcome next time is going to be five.

Finally, buried within a long-term trend, there might well be short periods where the data moves in the opposite direction. Some climate-change deniers liked to cherry-pick the period 2001–13 to 'prove' that global warming had stopped. Over that time-frame, the average global temperatures moved up and down with no obvious trend. Zoom out to look at the data over a century, however, and the evidence of an upward trend is compelling. That in itself is not proof, of course. But most scientists prefer to look at the long-term statistics, not short-term blips.

PREMIER LEAGUE GOALS: CERTAINTY
WITHIN UNCERTAINTY

Here's a prediction. Next season, 1,000 goals will be scored in the Premier League.

OK, it will probably be a handful more than that, but that figure is probably right to within 5%, which is astonishingly accurate by back-of-envelope standards.

How come this prediction is such a conveniently round number? Mainly it's a coincidence, but our confidence in it comes from history. If you look at all the seasons since 1995/6 (when the league settled at 20 teams), the highest number of goals scored was 1,066 in 2011/12, and the lowest was 931 in 2006/7. In six of the nine seasons between 2009 and 2018, the total number of goals was in the range 1,052 to 1,066.

In total there are 380 Premier League matches played in a season, with an average of about 2.6 goals per match. This means that were the league for some reason to change the number of teams up or down, we could still make a decent estimate of how many goals there would be. Increase the league from 20 to 22 teams, and we'd now have 462 matches. That's roughly a 20% increase in the number of games, so we'd expect about 1,200 goals.

If we increase a league to 24 teams, there are 552 matches; that's almost a 50% increase on the Premier League, so we might expect, say 1,450 goals. As it happens, the second division, known as The Championship, does have 24 teams. And sure enough, the average number of goals per season is roughly 1,450.

It might seem remarkable that a game that is popular for its drama and unpredictability can be so surprisingly predictable when you look at the big picture – but that's not unusual in statistics.

4

FIGURING WITH FERMI

THE FERMI APPROACH

The man often credited as the guru of back-of-envelope calculations was Enrico Fermi. Fermi was a physicist who was involved in the creation of the very first nuclear reactor. Most famously, he was present at the detonation of the first nuclear bomb, the so-called Trinity Test in New Mexico, USA, in July 1945. At the time, scientists still weren't sure how big the explosion would be. Some even feared it might be large enough to set off a chain reaction that would destroy the planet.

The story goes that Fermi and others were sheltering from the explosion in a bunker about six miles from ground zero. When the bomb went off, Fermi waited until the wind from the explosion reached the bunker. He stood up and released some confetti from his hand, and when it had landed, he paced out how far the confetti had travelled. He then used that information to make an estimate of the strength of the explosion. We don't know for certain how Fermi did this, but it probably involved him estimating the wind speed and working out how much energy was required to push out a 'hemisphere' of air from the centre of the explosion.

Fermi's estimate of the bomb's strength was 10 kilotons. Later, more rigorous calculations revealed that the real strength had been nearer to 18 kilotons, in other words Fermi's answer was out by a factor of nearly two. Anyone submitting an answer that far out in a maths exam would probably get no marks, yet Fermi got huge credit for the accuracy of his back-of-envelope answer. The important thing was that his answer was in the right order of magnitude, and gave scientists a much better understanding of the potential impact of the weapon they were now dealing with.

What was so impressive about Fermi's calculation was that he did it based on such crude data. His results also show that being in the right ballpark can sometimes mean being quite a long way from the exact right answer, and that an 'inaccurate' answer can still be useful.

Calculations that are done without access to much proper data have become known as 'Fermi problems'. These calculations are typically done as an intellectual exercise: working things out for the sake of it.

There is, however, a practical benefit to honing your skills at solving Fermi problems, because this type of problem is notoriously used in interviews. I can still remember one of the questions I was asked at my university entrance interview: 'What does an Egyptian pyramid weigh?'

One way to figure this out would be:

(1) estimate the dimensions of the pyramid and hence its volume;
(2) estimate the density of stone (in kilograms per cubic metre) and work out its volume;

(3) multiply the volume by the density to get a figure for the weight.

I doubt my interviewers were interested in the answer, not least because nobody has ever weighed an Egyptian pyramid. What the interviewers were looking for was the thinking process. I don't remember what answer I gave, but it can't have been too bad because I was offered a place.

For many employers it's the same: companies like Google and Microsoft are famed for the (Fermi) questions they have posed to would-be employees, to see how they think on their feet.

Whether you're practising for interviews, or are just exercising your brain for the fun of it, Fermi problems are an excellent work-out. In this section I have come up with a selection of Fermi problems that captured my imagination. In each case, I've shown the approach I would take when tackling them. You might well take a different approach. The only thing you and I can guarantee is that we are very unlikely to come up with the same answer, though with any luck we will at least both end up in the same ballpark.

COUNTING

When embarking on Fermi-style estimations, a good place to start is counting. Answering the question 'How many . . .?' is the most primitive mathematical challenge – though it often turns out to require a surprising level of skill, and is sometimes highly contentious too.

Who can forget Donald Trump's anger when his own view that he had 'probably the biggest crowds ever' at his inauguration was contradicted by several sources who suggested the crowd was a lot smaller than his predecessor's?

We've already seen examples of where even meticulous attempts to count exactly the right number often miss the mark (see vote-counting in elections on page 18, for example). Fortunately, there are many situations where only a decent estimate is required. Here are some examples.

HOW MANY WORDS ARE THERE IN YOUR FAVOURITE BOOK?

Publishers like to do word counts – sometimes they even pay a rate per word. So how many words are there in a manuscript, or a book you pluck off the shelf?

There is, of course, in most word-processing software, a button that can be clicked to find the precise answer to this question: but that's only for the electronic version of the book. For the paper version, you can either laboriously count each word or – more practically – make an estimate.

Word counting lends itself to the technique of taking a

sample and then extrapolating that to estimate the whole book: word length and density tends to be pretty consistent through most books, so if you turn to a random page of full text somewhere in the middle of the book, that's likely to be a good representation of every page.

How many words are there in, say, *Pride and Prejudice*?

The ultimate cavalier approach is to count the words in one line and base the entire estimate on that. Twelve words in the line, 38 lines on the page, 345 pages, using Zequals that's

$$10 \times 40 \times 300 = 120,000 \approx 100,000 \ words.$$

But it's not much effort to take a bigger sample and come up with a more accurate estimate. Even counting three lines of text to get an average will improve the estimate considerably. If the word count over three lines is 34, that's an average of about 11 words per line. And with the indents, sentences that end in the middle of the page, and half-pages at the end of the forty-odd chapters, we might say there's the equivalent of about 36 full lines of text per page, and about 320 pages.

The approximate answer hasn't changed: $11 \times 36 \times 320 \approx 10 \times 40 \times 300 = 120,000 \approx 100,000$. But these better estimates of words per page and number of pages justify a more precise estimate: $11 \times 36 \sim 400$, and $400 \times 320 = 128,000$.

Impressively, but perhaps fortuitously, it turns out that this estimate is remarkably close to the official word count of 122,000.

Working out the word count in the book you are reading now is a bit trickier – tables, numbers, diagrams and in-fill boxes make things a bit more complicated. But you can still make reasonable estimates to come up with a figure.

HOW MANY HAIRS ARE THERE ON
AN ADULT HUMAN'S HEAD?

The number of hairs on a person's head will, of course, vary hugely from person to person. Let's try to figure out a number for somebody with a full head of hair, say. And let's do it just using the imagination, and without looking at a scalp.

Picture (if you can) a square centimetre of scalp on a human head.

How far apart are the follicles? We can start by putting some upper and lower bounds on it. If hairs were 2 mm apart then the hair wouldn't be dense at all, and you'd think we'd be able to see the scalp shining through. But if hairs are as little as 0.5 mm apart, the scalp would be so densely packed, it would be more like fur. So let's go for 1 mm as a reasonable compromise.

That means that within each square centimetre of scalp, we're looking at $10 \times 10 = 100$ hair follicles.

What's the surface area of a scalp? Put your hands around your head, fingers at the top of your forehead, thumbs on your neck at the base of the scalp – call that a circle maybe up to 25 cm (10 inches) across.

The area of a circle with a diameter of 25 cm is: $\pi \times$ radius^2, and the radius in this case is 12.5 cm. But this is all so rough and ready that Zequals is called for:

$$3.14 \times 12.5^2 \approx 3 \times 10^2 = 300 \ cm^2.$$

The surface area of the scalp isn't a circle, it's more like a hemisphere. For the same radius there will be more surface area on a hemisphere than on a circle, so let's double it to 600 cm^2.

The total hairs on the typical head is therefore, according to this estimate, $600 \times 100 = 60,000$.

That will vary a lot – the number probably going up to 100,000 at most, and down to 30,000 at least (if a receding hairline or thinning hasn't set in).

The knowledge that a person has, at most, 100,000 hairs on their head means you can state with absolute certainty that there are at least two people in (say) Huddersfield[23] who have exactly the same number of hairs on their head as each other.

The proof is found by first supposing that everyone in Huddersfield has a *different* number of hairs on their head. We know that the most hairs anyone has is going to be around 100,000. So let's suppose that everyone else has fewer hairs, and that they all have a different number of hairs on their heads. Imagine now lining them up, starting with the person with no hair, then the one with one hair, two hairs etc.

We have 100,000 people with different numbers of hairs on their head. But what do we do with person 100,001? Or indeed with the other 50,000 or more Huddersfield residents? Inevitably, they are going to match with somebody. Hence we have proved our assertion that at least two people in Huddersfield must have the same number of hairs as each other, and indeed that there will be tens of thousands of people whose hair numbers are not unique. This form of proof is known as the pigeonhole principle, and is often used by mathematicans in problems rather more abstract than counting hairs.

---

[23] Any other town with a population well over 100,000 will do instead.

DO MORE PEOPLE GO TO FOOTBALL MATCHES AT THE WEEKEND THAN GO TO CHURCH?

Which is Britain's biggest religion, Christianity or football? It's a question that crops up periodically, especially when new statistics are published showing the decline in church attendance. It's a hard comparison to make, because quantifying church attendance requires a lot of sweeping assumptions and definitions, including what counts as a 'church', and what counts as 'going to church'.

Do weddings and funerals count, for example? And how many church weddings and funerals are there?

Wedding and funeral attendances are an interesting thing to estimate in their own right. There are around 250,000 weddings per year (see where this estimate comes from on page 198). I also know that these days most weddings don't happen in church. Let's suppose one-third of them do. That means maybe 80,000 church weddings per year. Let's say there's an average of 50 people at a church wedding. That suggests maybe:

80,000 church weddings per year

× 50 per wedding ÷ 52 weeks

≈ 80,000 people go to a church wedding each week.

That's not much higher than the number attending a Manchester United game at Old Trafford.

Funerals are more significant: if the UK population is roughly stable and is evenly spread across all age groups, then we'd expect a cohort of 900,000 to die each year. However, most deaths are still among those in the pre-Baby Boomer generations when the population was much smaller, which is one reason why the annual deaths toll in recent years has been nearer 500,000. But everyone who dies has a funeral, and church funerals are still the most common. So that suggests at least 250,000 church funerals per year, or 5,000 per week, with an average attendance of, I don't know, maybe 50 people? That would mean a substantial 250,000 funeral attenders in church each week.

All in all, weddings and funerals account for perhaps 350,000 church attendances per week.

Whether we ignore weddings and funerals or not, any official statement of church attendance has to be taken with a pinch of salt. In 2018, the Church of England claimed that an average of 750,000 people attended a church service each week, but unlike in a football match, there is no turnstile or pre-paid seating from which a headcount can be determined. There is a tradition in some churches of doing an annual census known as the 'October count'. But sources tell me that these figures include the occasional vicar scanning the congregation, putting her finger in the air, adding an optimistic 20% and saying 'I reckon we had 40 in this week, verger.'

Football crowds are much easier to count, as are the number of matches.[24] Over a weekend, 10 premier games (35,000 per game?), 12 Championship matches (15,000?), and 24 in Leagues 1 and 2 (10,000?) gives us about 750,000.

---

[24] Though see page 12 for why published attendances at football matches overstate the real figure.

Of course there are hundreds of other matches too, and rather than count them, we can probably apply something like the 80:20 rule here[25] and say that the rest of the matches are going to have a headcount of no more than 200,000.

So that's around a million people attending a football match in England in a typical week – that's more than the number of people going to a Church of England service. But by the time we add in Catholics, Methodists, Baptists, Pentecostalists and other denominations, that number of Christian church attendees can probably be doubled. So the Church undoubtedly has the edge over football. For the time being, at least.

HOW MANY TENNIS BALLS ARE USED AT WIMBLEDON?

Some employers have a reputation for asking curve-ball questions during job interviews, to see how well candidates can think on their feet. The Wimbledon tennis-ball question is attributed to the international consultancy firm Accenture, though it is not clear if this has ever actually been posed in an interview or is just an urban myth. Would an employer really ask a question that requires a reasonable knowledge of tennis? Well, maybe. And in any case, it's an interesting Fermi question.

---

[25] The 80:20 rule, more formally known as the Pareto principle, is a rule of thumb that suggests that 20% of the population owns 80% of the resources. It applies – very crudely – to many situations, including the distribution of wealth in countries and, very likely, to the attendances at football matches.

Let's assume we're talking purely about Wimbledon fortnight – the men's and women's singles, men's and women's doubles, and the mixed doubles.

First of all, how many matches are there?

Wimbledon is a knockout tournament, and once the main tournament starts, there are no byes. As with all such knockout tournaments, this means that the number of contestants must be a power of 2: there are 2 players in the final, 4 in the semi-final, 8 in the quarters, and then 16, 32, 64, 128 as you go back through to the first round.

There's a short cut for working out how many matches there will be in a knockout tournament. Since each match knocks out one player, and the tournament finishes with only one player who hasn't been knocked out, the number of matches must always be one less than the total number who enter the tournament.

There are 128 players at the start of a Wimbledon singles tournament, so the number of matches is $128 - 1 = 127$.

In the men's and women's doubles championships there are 64 pairs in the first round, and so 63 matches in total, and there are 48 in the mixed doubles, so 47 matches.

That makes the total number of main-event matches at Wimbledon:

127 + 127 + 63 + 63 + 47 . . . let's just call it 400.

Men's Wimbledon matches can last anything between three and five sets (call it four on average), while matches involving women will be two or three sets (call it two and a half on average). Combining the two, let's say an average Wimbledon match has three sets.

The number of games in a set can range from a minimum of six (when the set is 6–0) to a maximum of 13 when it goes to tie-break (7–6). That's an average of 9½ games, but let's round it up to 10, because that's easier, and also because the final set doesn't have a tie-break at 6–6, which pushes up the average number of games.

So an average Wimbledon match has about:

10 games per set × 3 sets per match

= 30 games per match.

One bit of technical knowledge that the apocryphal Accenture interviewers surely couldn't expect anything but a tennis nut to know is that the umpire asks for a set of six new balls ('New balls please!') every nine games. In other words, our tennis match lasting 30 games is going to require three or four sets of six balls – let's say 20 balls per match. Four hundred matches.

20 balls per match × 400 matches

= 8,000 balls.

That's a lot of balls, but this estimate is way below the 50,000 that Wimbledon claim to use each year. How come? For a start, there's the wheelchair, junior and celebrity tournaments that also go on during the fortnight – so you can probably double our 8,000 just with that. Then there's balls that players use for warming up, plus the balls that get hit into the crowd and get kept as souvenirs . . . and no doubt Wimbledon keep a few spares in the shop as well.

Put it this way: if you are in a job interview and you are asked how many balls are needed for Wimbledon, and if you use the reasoning above to get you to somewhere in the 5,000 to 50,000 range, you'll probably get a big tick in the assessment sheet.

HOW MANY ATTENDED DONALD TRUMP'S INAUGURATION?

Crowd counting can be very political. When there is an organised protest and march (be that for more pay, lower taxes or better rights for animals) the organisers want to be able to claim as big a number as possible, while the thing or people being protested against – often the government – would much prefer that number to be as small as possible. The police, acting on behalf of government, have a tendency to downplay the numbers too. As a result, you can almost guarantee that the size of crowds claimed by the protestors and protestees will always be very different, sometimes by a factor of two, depending on who you ask.

For example, there was a huge protest in London when Donald Trump made his first visit to the UK. The protestors were quick to claim that 250,000 were there. The police, according to the *Independent*, were reluctant to give a figure, but were prepared to agree that 'more than 100,000' had turned out.

Which raises the question: did more people turn out to protest against Trump in London than turned out to celebrate him at his inauguration in 2017?

One person whose answer would be no is, of course,

Donald Trump, who was adamant that there had never been a bigger crowd. As far as (most of) the rest of us could tell, however, Trump's inauguration crowd was a fraction of that of his predecessor, Barack Obama.

But how big was it?

One unreliable way to estimate a crowd size is to view it from a low angle, for example when speaking to it from a raised podium. Looked at this way, even partially filled space can look full because the floor is not visible. Given that this is the view that Trump had of his crowd, you can understand his over-estimate. Well, maybe.

Professional crowd estimators prefer to view the crowd from above, from a photo taken by a drone, for example. This way it's possible to see how packed the crowd is. Common practice is to divide the photo into squares of (say) five metres, look at the crowd density within each square, then group the squares into categories from 'packed' to almost empty.

There are commonly used guideline figures for a crowd's density, based on the number of people per square metre:

No. per square metre	Nature of outdoor crowd
1	Blanket/picnic crowd (e.g. outdoor classical concert).
2	Crowd on Henman Hill at Wimbledon.
3	Political rally.
4	By the barriers as the Queen drives past.
5–6	Fans at the front of a One Direction concert.
>6	An uncomfortable crush, individuals likely to be pressed up against each other.

But for Trump's crowd we don't have the luxury of aerial photos, so our estimate is going to be very crude. Here's a sketch of the National Mall:

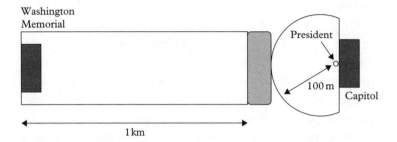

Since Trump's crowd by no means filled the National Mall in Washington DC, we can assume that people had no need to pack themselves to maximum density. It was very full in the big semi-circular section in front of the Washington Capitol, and the crowd also stretched back into the straight section between the Reflecting Pool and the Washington Monument.

The semi-circular section is about 100 metres radius, so its area is roughly:

$$\pi \times 100^2 \div 2 \approx 15{,}000 \text{ m}^2.$$

That section was very full, and therefore probably comfortably packed, so let's assume its density was three people per square metre. That means a crowd of around 50,000 (a decent Premier League crowd) immediately in front of Trump.

The straight Mall beyond the lake was not packed – the event planners had put down white flooring, so it was easy to pick out the empty sections. The straight is around 1 km (1,000 metres) long, and the covered segment perhaps 100 metres wide. So the total area of the straight is around:

$$100 \times 1,000 = 100,000 \ m^2.$$

Let's say the whole segment was 50% occupied by spectators (that's generous), and the space was comfortably full in those sections, i.e. two people per m^2. We're looking at 200,000 in the straight section, and 50,000 near the front. That's maybe as many as 250,000 in total? (Interestingly, 250,000 is the figure that one of the expert sources gave as their estimate as well.)

Since 250,000 is the upper limit of the London protest crowd, it's likely that there were more at Trump's inauguration, but it seems that it was a fairly close-run thing.

WHAT ARE THE CHANCES?

People love coincidences. We are drawn to them instinctively, because the human brain is geared to look out for patterns. When we observe something surprising, we want an explanation, and we assume that something has caused it. If there is no obvious physical cause (such as somebody fiddling the lottery) then we reach out for explanations that are supernatural.

And the hunt for an explanation inevitably includes the question: 'what are the chances?' – the more remote the odds, the more excited everyone gets.

But how are these odds worked out? If ever there are situations that call for the back of an envelope, it is these. Let's look at some coincidences in more detail.

WHAT'S THE CHANCE OF WINNING THE LOTTERY TWICE?

On 10 September 2009, the Bulgarian Lottery hit the headlines. Four days earlier, the six numbers drawn had been 4, 15, 23, 24, 35 and 42. Now, in the very next draw, exactly the same six numbers were pulled out. 'What are the chances?' asked almost everybody (to which the answer reported in the media was 'about one in five million'[26]).

---

[26] There are conflicting reports of how many balls were in the Bulgarian lottery. The five million figure is on the basis that there were 42 balls in the lottery. Other sources suggest there were 49 balls in the Bulgarian lottery in question, in which case the odds of a repeat of the numbers was 14 million to one.

With some coincidences that we'll see later, all sorts of hand-waving assumptions are needed before you can come up with an estimate of the probability. Not so the result of a single lottery. These games are carefully engineered so that the outcomes are entirely random. And the conditions of the game are well defined: there are a fixed number of balls (typically somewhere between 40 and 60 in most lotteries), each of which has exactly the same chance of being drawn.

This means that any combination of six numbers is equally likely: next week's numbers in the UK Lottery could be 7, 12, 14, 23, 41 and 58. Or, just as likely, they could be 1, 2, 3, 4, 5 and 6.

If the numbers 1, 2, 3, 4, 5 and 6 were to be drawn, it would be headline news of course, not because this combination is any less likely than any other, but because it makes a far more interesting pattern.

So the reports of the freak occurrence in the Bulgarian Lottery, where the identical numbers came up in consecutive draws, were right. This was indeed a 'millions to one against' occurrence – and we haven't had to do any estimation.

However, there are other lottery coincidences that *do* require some back-of-envelope thinking.

In June 2018, an anonymous Frenchman won a million euros in his country's lottery for the second time in two years. The odds against this were quoted in headlines as being 16 trillion to one, that's 16,000,000,000,000:1 written long-hand. When numbers get this big, we have little sense of what the numbers mean – millions, billions, trillions, they might as well be called gazillions; they all sound the same. Sixteen trillion is more than two thousand times the number of people on the planet. It sounds unlikely as a figure. In fact, it is so big, it sounds suspicious.

Where did '16 trillion' come from?

We do know that the chance of winning the French My Millions Lotto in any given draw is around 20 million to 1. If you want to work out the chance of *two* events happening, such as tossing heads on a coin and rolling a six on a dice, then, so long as the events are completely independent, you simply multiply the probabilities together (see page 106). The heads and six combination has a probability of $\frac{1}{2} \times \frac{1}{6} = \frac{1}{12}$. This same maths applies to a lottery. The chance of entering the French Lottery exactly twice and winning both times is:

$$\frac{1}{20} \text{ million} \times \frac{1}{20} \text{ million,}$$

. . . which works out at 1 in 400 trillion – far longer odds than the 16 trillion we saw above.

But our French winner didn't just play twice and win both times. He won twice over a period of 18 months, and it's likely that he played the lottery many times during that period. In fact, like most lottery participants, he probably played every week. This is going to significantly reduce the odds, simply because he now had a lot more opportunities for a double win.

To work out the odds of this, we need to start making some guesses.

Let's suppose he played Lotto every week. This means he played about 75 times over the 18 months. Looking at all the possible combinations of dates, it works out that there were nearly 3,000 different pairs of dates[27] on which the Frenchman could have won the jackpot over a period of 18 months. This means the odds were roughly:

$$\frac{1 \times 3,000}{400 \text{ trillion}},$$

reducing the odds to around 130 billion to one.

That's now far smaller than the 16 trillion figure. Perhaps the academic who came up with the number 16 trillion was only looking at ways of winning the lottery with a gap of exactly 18 months 'plus or minus a few weeks'. In other words, he was saying 'What is the chance that somebody would win the lottery, and then win it again *about a year and a half later*?' And if you set the level of vagueness right, the answer to that question might legitimately be '16 trillion to one' . . . but it's a pretty meaningless figure.

But there's another huge flaw in quoting the odds as being in the trillions. Yes, that was the chance that this particular monsieur would win the lottery twice in the stated period. But before he won the first time, we had no interest in him as an individual, because we know with 100% certainty that *somebody* was going to win the Lotto. Only when somebody is already a winner are we interested in whether they win again. In other words, rather than asking 'What is the

---

[27] The number of different pairs of dates that can be chosen out of a list of N dates is $\frac{1}{2} \times N \times (N-1)$. In this case $N = 75$ and the total number of pairs is 2,775

chance of a particular person winning the lottery twice in two years?', we should be asking: 'What is the chance that somebody who has won the lottery once will win it again inside two years?'

Remember that the chance of winning the French Lottery is about 1 in 20 million. If a person who wins plays once a week, they have roughly 100 chances to win the lottery again in the two-year period, so the odds become 20 million divided by 100, or 200,000 to one. Still a big figure, but tiny compared to the 16 trillion we were first presented with.

And that's if they only buy a single ticket each week. Lottery winners have so much money sloshing around, they can easily afford to buy 100 tickets a week and not notice the cost. Maybe that's what our Frenchman did – we have no idea, since he kept himself secret. In that case, we would be looking at odds in the low thousands, not the millions.

This mistake of including the first instance of an event in the calculations helps to make the odds sound more impressive, and therefore newsworthy. That's why newspapers invariably do it when reporting an interesting coincidence.

It reminds me of the story of the man who was terrified of bombs on planes. On his next flight, he tried to take a bomb onto the plane with him. 'What the hell do you think you're doing?' asked the security officer. 'Well,' replied the man, 'I've heard that the chance of there being a bomb on a plane is about one in a million. So I thought I'd take on a bomb myself, because I've figured out that the chance that there will be two bombs on the plane will be one in a trillion.'

THE EXACT ODDS OF WINNING A LOTTERY

The odds of winning the jackpot in any lottery can be worked out precisely. In the UK, where six balls are drawn from 59, the odds of any particular combination of six balls being selected is worked out by calculating the total number of ways in which six balls can be drawn from 59. The calculation looks like this:

$$\frac{59!}{53! \times 6!}$$

Here, 59! means the product of all the numbers from 1 to 59, otherwise known as 59 factorial. Written out long-hand (skipping the middle numbers) it looks like this:

$$\frac{59 \times 58 \times 57 \times 56 \times 55 \times 54 \times 53 \times 52 \times \ldots \times 3 \times 2 \times 1}{53 \times 52 \times 51 \times \ldots \times 3 \times 2 \times 1 \times 6 \times 5 \times 4 \times 3 \times 2 \times 1}$$

The 53! can be cancelled out from top and bottom and the calculation simplifies to:

$$\frac{59 \times 58 \times 57 \times 56 \times 55 \times 54}{6 \times 5 \times 4 \times 3 \times 2 \times 1}$$

A quick estimate should convince you that this is going to work out to be a very large number. In fact, the (precise) figure is: 45,057,474.

In other words, if you pick six numbers at random, there's roughly a 1 in 45 million chance that they will come up in the next draw.

WHAT WERE THE CHANCES OF THE ORKNEY BABIES?

On 13 November 2018, two women – Angela Johnston and Karen Daily – both gave birth in Balfour Hospital in the north of Scotland. What made this more unusual was that both women came from Stronsay in the Orkney islands, an island with a population of 350, where babies are a relative rarity. Even more remarkably, the two women had both taken the same ferry to the mainland hospital a few days earlier. Both had (quite independently) decided that if their baby was a boy, they would call it Alexander. And then – to the amazement of everyone – both babies arrived at 11:36 p.m.

'What were the chances?', asked everyone, including BBC Radio Scotland, who then rang me for the answer. I guess they were assuming that this is the sort thing that does have 'an answer' one can just rattle off.

But as with all the other estimates in this book, we have to make some assumptions before we can do any meaningful calculations.

First of all, how many babies would we expect to be born in Stronsay each year?

If a population is going to be stable, rather than growing or declining, the number of births and deaths per year has to be roughly the same. For a population with a regular age distribution, with life expectancy of 80 years, we might expect

⅟₈₀th of the population to die each year. The population of Stronsay is 350, so this first stab at an estimate suggests we might expect

Number of births per year = 350 × ⅟₈₀ ≈ 4.

Of course I may be wrong to assume that Stronsay's population level is stable, and even if it is, those who die might be replaced by migrants rather than by babies. Still, it seems reasonable to suppose that an island like Stronsay might expect two babies each year. Let's suppose it's *exactly* two babies per year.

What's the chance that both those babies will arrive in the same minute?

The number of minutes in a year is $365 \times 24 \times 60$.

We can use Zequals to work this out. The number of minutes in a year:

$$≈ 400 \times 20 \times 60 ≈ 500,000,$$

which is half a million. So, if my assumptions are correct, then the chance of a pregnant mother giving birth in a particular specified minute in the year is about one in half a million.

However, remember that the first baby was inevitably going to be born at some minute. To work out the chance that both babies will be born in the same minute, we only need to consider the second baby (just as we did with the lottery winner and his second jackpot). The chance that the second baby would arrive in the same minute was about one in half a million, as calculated above. So that's my first pass at an answer.

But what about those other factors? Catching the same ferry to the same hospital? Actually, there is little surprise here: ferries to the mainland probably don't run that often, and if the mums' due dates were similar then it's not a huge surprise they caught the same ferry to the nearest hospital that has a maternity unit.

Then again, what about those names? The chance they'd both call their child Alexander was much more of a coincidence. According to a recent census, perhaps as few as 1 in 100 boys are given the name Alexander[28] in Scotland. So this would have carried the odds into lottery-winning territory.

There was just one catch. Mrs Daily gave birth to . . . a baby girl. There was no second Alexander after all.

WHAT'S THE CHANCE OF TWO HOLES IN ONE?

In October 2017, Jayne Mattey and Clair Shine were enjoying a round of golf in Berkshire. It was hole 13, a number viewed as unlucky for some, but not for Jayne and Clair. Jayne teed off first and to her amazement the ball hit the pin and dropped into the hole. It was the first hole in one of her life. Then Clair took aim: the ball went straight, and to their astonishment, it too rolled in. The chances? According to the National Hole-in-One Registry, which is based in

---

[28] According to the National Records of Scotland, in 2018 there were 275 babies named Alexander out of a total of 24,532 baby boys born in the year.

North Carolina, USA but likes to keep tabs of golfing feats across the world, it's '17 million to one'.

Those are very similar to the sort of odds that are quoted for winning a lottery.

But there is a big difference between holes in one and the lottery, because, as we've seen earlier, while the odds in a lottery are fixed and can be worked out exactly, the odds of a hole in one are more about sticking a finger in the air and making a lot of assumptions.

For a start, those odds depend on the player. A top golfer such as Rory McIlroy or Tiger Woods is far more likely to hit the pin than a regular club player.

Then there's the length of the hole. In order to get a hole in one, you need to be capable of hitting the ball from the tee to the green. For almost all golfers this means holes in one can only happen on shorter holes where a decent player is expected to take three strokes to get the ball into the hole (these are known as 'Par 3 holes'). These holes are typically between 100 and 200 metres long. The shorter the hole, the easier it is to get a hole in one because an error in the direction that you aim the ball is not punished so severely.

There are usually four Par 3 holes on a golf course, so in a round of 18 holes, there are four opportunities to get a hole in one. This means that the chance of getting a hole in one during a round of golf is about four times higher than the chance of achieving it on a particular, named hole.

The 'One in 17 million' figure came from the National Hole-in-One register. Based on statistics that they have accumulated from around the world, they reckon that the chance of getting a hole in one on a specific, named hole is around 1 in 2,500 for a professional golfer, and about 1 in

12,000 for a club golfer. So, on a given hole, we might expect the chance of two average women golfers both getting a hole in one to be:

$$\frac{1}{12,000} \times \frac{1}{12,000}.$$

That's more like 1 in 150 million.

But further investigation into the story reveals that the hole in question had been shortened to just 90 yards because the course was being repaired, which must have hugely reduced the odds of a hole in one. And also, the women were playing in a group of four, whom we will call A, B, C and D. This means there were six possible pairs of women who could have got a hole in one: AB, AC, AD, BC, BD and CD. And it would have been a headline if any of those six pairs had been successful, so we can divide the odds by six. All of which brings our 1 in 150 million down to something much smaller.

In the end, nobody really cares if the odds are 1 in 17 million or 1 in 50 million: it's merely a chance to report that something freakishly unlikely just happened. But was it really freakishly unlikely? Before I went in to record a radio item about this story,[29] I thought I'd drop into my nearest golf club, in Dulwich in south London, to see if I could pick up any anecdotes. I spoke to the manager.

'Holes in one? We get about ten of them a year. In fact, we had an eleven-year-old get one last Sunday,' he said, and he showed me the card sitting on his desk.

---

[29] It was a BBC World Service edition of *More or Less* in January 2018 – you'll find it online.

'Oh, but if this is about those two ladies who got holes in one last week, we can beat that.' He led me to a plaque on the wall outside his office. With a caption 'Halved with holes in one!' it showed two smiling men who had just tied in a match-play tournament with holes in one. This had happened in 1984.

So, the first random golf club I turned up at was able to produce an equally freakish story to the one experienced by the ladies foursome in 2017.

I made a quick back-of-envelope calculation:

No. rounds played in Dulwich per year 30,000

No. Par 3 holes played per year 30,000 × 4 ~ 100,000

No. Par 3 holes in 30 years 30 × 100,000 = 3 million.

In other words, in 30 years there have been about three million opportunities for two people to get a hole in one at the same time, and it has happened at least once. It confirms that odds of this event are probably one in a few million.

However, since there are estimated to be over 500 million rounds of golf played around the world each year, we'd expect the story of the two ladies and the holes in one to be replicated several times each year.

And sure enough, this turned out to be the case.

To find more details about the Berkshire ladies hole-in-one story, I searched online using the phrase 'Two women golf hole in one'. The first story to pop up was not about the Berkshire ladies. It was an almost identical story from Northern Ireland that had happened a couple of months earlier. This time it was Julie McKee and Mandy Higgins who had got holes in one, again as part of a foursome. Their fluke was described as a one-in-a-million story. Which just goes to show how hand-wavy these headline odds can be.

WHAT'S THE CHANCE OF DEALING OUT FOUR 'PERFECT' HANDS?

We've seen some remarkable coincidences already, but on the face of it, one of the most remarkable coincidences of all time happened in Kineton, Warwickshire, in April 2011. Four Warwickshire pensioners were playing a game of *whist*, a traditional card game in which the entire pack of 52 cards is dealt out between four players. Each player receives a hand of 13 cards.

The pack was shuffled and then dealt. To their utter astonishment, when they picked up their cards, each of the four players discovered they had been dealt a complete suit.

One of the pensioners, Wenda Douthwaite – who was dealt a hand containing all 13 spades – declared herself to have been 'gobsmacked'.

141

'I've never seen anything like it before', she said. And Wenda's surprise was justified, because mathematically, the chance of this happening in a random deal of a regular, thoroughly shuffled pack of cards is a staggering one in 2,235,197,406,895,366,368,301,559,999.[30] In other words, there are about two *octillion* different combinations of cards that can be dealt out into four equal hands, and in only one of those does each player get a complete suit.

Let's just look at how unlikely this was to happen.

There are currently over seven billion people on the planet. Imagine giving each of them a pack of cards, and getting them to shuffle it thoroughly and deal out four hands.

Now get them to repeat that every minute, so 60 deals per hour. Let's allow them nine hours in the day to sleep and eat. That leaves them the remaining 15 hours a day to deal cards.

They can deal out:

$$15 \times 60 = 900 \text{ deals per person per day.}$$

So the total number of deals in a year would be:

$$900 \times 7 \text{ billion} \times 365$$
$$\approx 2 \text{ quadrillion (that's } 2 \times 10^{15}\text{) deals per year.}$$

Even if every hand dealt was different, to deal all of the two octillion possible sets of four whist hands would therefore take:

---

[30] As calculated by Peter Rowlett in *Aperiodical*, November 2013.

$$\frac{2 \times 10^{27}}{2 \times 10^{15}} \approx 10^{12} = \textit{one trillion years.}$$

Scientists reckon that the solar system will come to an end eight billion years from now, and so we can safely say that, according to those odds, not only is it unlikely that this hand has ever been dealt before in the history of card-playing, but even if everyone on the planet were to shuffle and deal a pack of cards every five minutes, it is unlikely to ever happen again before the universe comes to an end.

Which is very curious, because in 1998, in Bucklesham, Suffolk, four pensioners playing a game of bridge experienced exactly the same coincidence. Hilda Golding, one of the players involved on that occasion, said at the time: 'I was amazed. I'd never seen anything like it before, and I've been playing for about 40-odd years.'

And if you trawl through the archives, you'll discover other reports of the same thing: in Pennsylvania in March 1938, Virginia in July 1949, Wyoming in April 1963 and many, many more. Among the reports of four players being dealt perfect hands is one from St James' Club in London, in 1959 . . . on 1 April. That date rings some alarm bells.

The point is that each of these occurrences was supposedly a once-in-the history of the universe event. The numbers do not make sense: the chance of all of these coincidences happening was so small that we are justified in calling them impossible.

Since the figures do not make sense, there must be another explanation as to what was happening.

There are two possibilities.

The first is that the cards being dealt had not been completely randomly shuffled. A new pack of cards comes with the cards neatly ordered by suit – all the spades, then all the hearts, the clubs and the diamonds. If you cut the pack exactly in half, and do two perfect riffle shuffles, so that the two halves of the pack perfectly interweave with each other, then the pack will now be arranged in the order: spades, clubs, hearts, diamonds, spades, clubs, hearts, diamonds and so on, all the way through the pack. If you now deal the cards out to four people, the first player will get all the spades, the second will get all the clubs, and so on. I'm not saying this is what happened. But it might have done.

Even when the cards have been shuffled, when they are gathered together at the end of a hand, they will tend to be grouped together by suit.

It takes at least seven good (but not perfect) riffle shuffles to be reasonably confident that the cards have been fully mixed up, and even then, they may well retain some trace of a pattern (spades clumped together, for example).

So in all of the stories of the freak card deals, it is extremely likely that after shuffling, the cards being dealt were not in a 'random' order at all. And while a perfect deal would still be extremely rare, the chances of it happening when the cards have a trace of order in them is astronomically higher than if they are completely mixed.

And there is a second possible explanation. How hard would it be for a practical joker to arrange the cards without the players knowing about it? A magician could do it easily – by switching the pack for another while using a diversion to distract the attention of the players. It would be even easier if one of the players were themselves the practical joker. And when the coincidence crops up on 1 April at a gentlemen's club, the chance that some underhand activity was behind it strikes me as really quite high.

There is something of a paradox here. The more unlikely a coincidence is, the more reason we have to believe that all is not as it seems. Imagine tossing a coin and getting heads 10 times in a row. That would be surprising and a bit unnerving. The chances of it happening are:

$$\frac{1}{2} \times \frac{1}{2} \times \frac{1}{2} \times \frac{1}{2} \times \frac{1}{2} \times \frac{1}{2} \times \frac{1}{2} \times \frac{1}{2} \times \frac{1}{2} \times \frac{1}{2}$$
$$= \left(\frac{1}{2}\right)^{10} \approx 1 \text{ in } 1{,}000.$$

But suppose you keep flipping that coin and it comes up heads another 90 times, meaning that you have now got 100 heads in a row. The chance of this happening at random is $\frac{1}{2}^{100}$, or one in a *million trillion trillion*. What's the chance that the next toss will be a head? Standard probability theory will tell you that the chance of another head is still $\frac{1}{2}$. But the astronomically remote odds of you getting 100 heads in a row up to now means that other scenarios come in to play. What's the chance that this coin is actually double-headed? Or that you always flip a coin identically so that it lands the same way up? Or that you have been hypnotised to believe that the coin is showing heads even when it isn't? All of these are unlikely, yet they are far more likely than the chance that it is a fair toss with a fair coin.

Put it this way – if I were to toss a coin 100 times and get heads every time, and you asked me: 'What are the chances that the next toss will also be a head?' my answer would be: 'It is almost certain.'

ENERGY, CLIMATE AND THE ENVIRONMENT

The future of the planet and how we treat it is one of the most pressing concerns of modern life. Nobody can be certain what the impact of climate change is going to be, as can be seen from the wide range of forecasts that come from the experts who build sophisticated computer models to predict this very thing.

The solutions on offer are consistent, however: reduce carbon dioxide and methane emissions (partly by saving energy), reduce the amount of waste that we produce, and for the waste that we can't avoid, try to recycle as much of it as we can. Back-of-envelope calculations can help us to get a sense of both the scale of the problems, and the priorities for how to fix them.

WHAT USES THE MOST ENERGY AROUND THE HOUSE?

We're being urged to do our bit to reduce global warming by saving energy. So it's time to think about making some personal cutbacks.

Imagine you're living alone in an apartment. You have a well-stocked fridge, a large TV on standby, you take a three-minute shower every morning, and you boil the kettle four times a day for coffee, tea and other essentials.

Which of these do you reckon is using the most energy over 24 hours?

(a) The fridge
(b) The TV on standby
(c) The shower
(d) The kettle

Among most audiences, there's one answer that tends to be more popular than the others: (b), the TV on standby. There are probably two different reasons why people choose this. The first is a recollection of hearing once that TVs on standby use a lot more electricity than you'd expect. The other reason people choose the TV is to second-guess the questioner ('the answer is going to be a surprise, so I'll choose that one').

In fact, the only one of the four options that is definitely *not* the biggest energy user is the TV standby. Many years ago TVs on standby were indeed quite energy hungry (the TV would get quite warm, which is where the energy went) but those days are gone. A TV on standby now typically uses 1 to 2 watts, a fraction of the power of a regular lightbulb.

As to the biggest energy user, well – it depends.

A regular fridge (with no freezer) typically uses about 50 W of power, not dissimilar to a normal lightbulb, but that does depend on how hard it is having to work (more energy is needed on hot days) and how efficient it is. It probably only has to work that hard for about half of the day, so that's:

$$50\,W \times 12\ hours$$
$$= 600\ watt\ hours.$$
$$\sim \tfrac{1}{2}\ kWh\ in\ a\ day.$$

A typical kettle uses 2 kW. If it takes three minutes (a twentieth of an hour) to boil the kettle, that's $\frac{1}{10}$ kWh per boiled kettle, so if we boil it four times, that's:

$$\frac{1}{10} \times 4 = \frac{4}{10} = \frac{2}{5} \; kWh \; in \; a \; day.$$

But if the kettle is full, it might take a lot longer to boil, and we could easily get up to over ½ kWh.

And the shower? Suppose you boiled four kettles and put that in a water tank, and topped it up with water from the cold tap so that the temperature was hot but not boiling. How long would that last you as a shower. A minute or two, perhaps? So the energy need for that hot shower is not far off the amount needed for the four boiled kettles.

Depending on how hot the weather is, how full the kettle is and how long the shower takes, any of the three appliances could end up being the biggest consumer. In terms of orders of magnitude, they can all be treated as the same.

But there's another everyday 'appliance' that on a typical day can consume energy at a higher order of magnitude. The car. If you start up your car and drive around town at a typical speed of 20 to 30 mph, stopping and starting at traffic lights, your rate of consumption of energy (or power) will typically average about 20 kW. In other words, driving your car is roughly the equivalent of switching on 10 kettles, and leaving them boiling for the entire duration of your journey. A 30-minute car journey consumes more energy than all your domestic appliances put together. That's something to think about when doing the school run. And as for that flight to Ibiza . . .

HOW MANY KETTLES IN A LIFETIME?

It's sobering to think about how disposable so many of the products we use every day have become. It's at least 10 years since two repair shops close to where I live shut down (one fixed televisions, the other fixed hoovers). Nowadays, when a TV or a vacuum cleaner stops working, we just chuck it out. And the sheer volume of 'stuff' that all of us might chuck out per year is mind-boggling.

Let's just pick one innocent household item as an example: the kettle that we looked at in the previous section. When I was a boy, we had an aluminium kettle that sat on the gas hob. We used that when making tea, and it lasted my entire childhood. But since I left home, I've always had an electric kettle, as most of us do. And kettles are the sort of thing you might expect to just be 'there' for a whole lifetime. But how many kettles is that likely to be?

I've been married for 20 years, and a kettle was one of our wedding gifts (do people still give kettles as wedding gifts?). Sadly, and please don't tell whoever gave it to us, that kettle only lasted a couple of years. We are now on our fifth kettle since the first one stopped working. Have we been unlucky, or does this mean that a typical kettle now lasts only three or four years? If we assume kettles are something adults are responsible for (and children are just beneficiaries of the kettles that are there), then it looks like somebody living to 80 might easily be getting through 20 kettles in their adult life. OK, so households usually have more than one adult, so that might be more like 20 kettles per household rather than per person, but still, my grandmother's generation would be staggered if they could see the rate at which we get through appliances that used to last most of a lifetime.

As a society we have become used to this pattern of short-lived, disposable household goods in the last 30 years or so, and it's now hard to imagine how life was beforehand. But let's look ahead 100 years. At this rate of disposal . . .

30 million households in the UK

100 years ÷ 4 years per kettle = 25 kettles per household

30 million × 25 = 750 million kettles

. . . we will have got through close to one *billion* more kettles by then. In the UK.

Suppose this does happen. If we're lucky, the metal elements from most of them will have been recycled, but the rest will be filling landfill sites somewhere.

Is this sustainable? Surely not. And what about in 500 years' time? It's hard for us to imagine what life will be like by then, but long beforehand there is going to have to be a seismic shift in our lifestyles and our consumption.

HOW MUCH FRESH WATER IS FLUSHED DOWN LONDON TOILETS EVERY DAY?

One of the unsung luxuries of modern living is that we have apparently limitless drinkable water available to us at all times. Yet a huge amount of that drinkable water, which is treated at some expense, is never consumed but is literally flushed away. But how much?

London has a daytime population of around 10 million, including those who travel to the city to work. We can be confident that all of them will use a toilet at least once, and more likely it'll be five or more times during the day. So we're looking at 50 million toilet visits in London per day.

Pondering for a moment (but not for too long) on practicalities, all 'female' visits are going to involve an individual flush, but perhaps only 25% of 'male' visits will do so (the majority of them being at urinals). So we're looking at, say:

25 million female flushes

25 × 0.25 male flushes

~ 30 million flushes per day.

How much water is sent down the drain during a single flush? It is going to vary from WC to WC,[31] but picture filling the tank using one-litre jugs, and even the most eco-friendly flush is going to use up three or four litres. So a very conservative

[31] In central London it varies from WC1 to WC2 of course.

estimate would say that London alone flushes away 100 million litres each day, and it's probably going to be a lot more.

A hundred million litres is 0.1 billion litres. To put that in perspective, an Olympic swimming pool holds about 2.5 million litres, so every day, toilet flushing in London could empty the equivalent of up to 100 swimming pools – the size of a small lake.

Of course water in itself is environmentally neutral. But the creation of reservoirs inevitably interferes with natural surroundings, because it means diverting rivers, building damns and pumping water. And in times of drought, water being used by humans is water that is being diverted from other fauna and flora.

COWS VERSUS HUMANS – WHICH EMITS THE MOST METHANE?

On the subject of effluent, let's turn our minds to cows. Why? Because of methane.

Methane is one of the worst greenhouse gases: in fact scientists estimate that, over a 20-year period, methane can trap 100 times more heat in the atmosphere than the same amount of carbon dioxide.

The alarming increase of methane in the atmosphere has largely been blamed on cows, especially beef cattle, whose methane emissions when digesting grass produce monumental quantities of the gas. (Contrary to popular belief, this is mainly through burping rather than flatulence.) That's why there's an urgent push to reduce the world's consumption of beef.

The average cow produces somewhere between 200 and 500 litres of methane per day (that's a huge figure, not one I felt qualified to estimate at all, so I looked it up – and even official sources vary hugely in the figure they quote).

Cows aren't the only creatures responsible for methane. Every living creature contributes methane as a natural part of its digestion or decomposition. That includes humans. In the seminal paper 'Investigation of Normal Flatus Production in Healthy Volunteers' by J. Tomlin, C. Lowis and N.W. Read (what do you mean, you haven't read it?), the authors concluded that the average human on a diet that includes 200 g of baked beans, produces about 15 ml of methane per day. To put that in context, remember that the figure for cows is of the order of hundreds of *litres* per day. So the average cow produces over one thousand times as much as the average human.

There are, of course, a lot more humans than cows (in the UK, seven times as many), but that still means that the total methane output of cows is hundreds of times higher than that of humans. And while the ratios will differ in other countries, it's reasonable to suppose that the global picture is similar: methane is mainly a cow problem.

Still, there are billions of humans, so how much is the world's human flatulence contributing to global methane levels?

15 mL × 8 billion ≈ 100 million litres per day.

That's roughly enough to fill London's Albert Hall, or the main concert hall of the Sydney Opera House, or – appropriately – one of the old Victorian gasometers that can still be seen dotted around the landscape of Britain.

HOW MANY PLANES ARE IN THE SKY AT THE MOMENT?

In 2016, the BBC ran an excellent three-part documentary called *City in the Sky* (hosted by Hannah Fry and Dallas Campbell) in which it was revealed that at any time, one million people are flying in a plane. It seems a staggering figure. Many thousands of planes in the air at any time, each spraying carbon dioxide high into the atmosphere. But can this figure be true?

Most of us probably notice at least a couple of planes flying overhead every day, but it is a stretch (for my imagination at least) to picture those few hundred people overhead representing one million constantly in the air across the world.

To start this estimation, you might think first about an extremely busy airport. The one I'm most familiar with is London's Gatwick Airport. Watching from the terminal I never have to wait long for a plane to take off: I'd guess one plane departs every minute. Those planes might spend anything between 30 minutes and 15 hours in the air, but what's the average duration of a flight? Maybe two hours?

So if each Gatwick plane is in the air for 120 minutes, and each minute one Gatwick plane is taking off, while one of the earlier flights is landing, that suggests there are about 120 Gatwick planes in the air at any moment. However, planes don't take off at this rate for all 24 hours of a day. Noise restrictions mean that airport activity is very limited during the small hours. So let's halve that figure to, say, 50 planes from airports like Gatwick are in the air at any time.

London Heathrow is busier than Gatwick, but the volume

of flights tails off when you think about other airports in the UK. Is it reasonable to guess that there's the equivalent of five Gatwick airports sending out planes in the UK? That would mean that 250 planes that set off from the UK are in the air at any time.

The number of planes leaving a country is likely to be linked to the size of its economy. Rich countries, and those with large populations, will no doubt fly more planes than small, poor ones. The USA is five times as populous as the UK, and richer. It's also spread over a bigger area, which will increase the need for planes. So if the UK has 250 planes in the air, then the USA surely has 3,000. Of the nearly 200 other countries in the world, we can probably ignore all but a handful – but combine the biggest economies, perhaps they are the equivalent of another 10 USAs? That would suggest something like 30,000 planes in the air at any time. If there are, on average, 50 people on a plane, we get:

50 passengers × 30,000 planes
= 1.5 million passengers in the sky,

which is consistent with that description of a 'city in the sky'.

The 'official' figures for air traffic published online suggest 30,000 planes in the sky is an over-estimate, with a variety of sources suggesting the real figure is somewhere between 5,000 and 10,000 (though this doesn't account for private and military planes). But even at the low end of the estimate we're talking about hundreds of thousands of people and millions of tons of metal in the sky above us. And a lot of fresh carbon dioxide being sprayed into the atmosphere, too.

CAN WE PLANT A TRILLION TREES?

According to the National Oceanic and Atmospheric Administration, the level of carbon dioxide in the atmosphere has increased by more than 25% in the last 70 years. It's been known for over a century that carbon dioxide in the atmosphere magnifies the greenhouse effect, and therefore contributes to global warming.

One way we might reduce the amount of carbon in the atmosphere is to plant trees, as they absorb the gas. By the time a tree has matured, it will have locked away anything up to a ton of carbon dioxide. But how many trees would it take to offset the carbon dioxide that is produced each year?

In 2017, a group of conservation organisations, including the World Wide Fund for Nature, launched a campaign called 'Trillion Trees'. The idea is to increase the number of trees across the globe by one trillion by the year 2050. This, they reckon, would be enough to compensate for the world's current CO_2 emissions. It is an ambitious and noble aim, though not everyone agrees that even this mass planting would be enough on its own.

But a *trillion* trees? How can we get our heads around that number?

One way to start is by thinking of some familiar woodland, or a planted forest. I grew up near Delamere Forest in Cheshire. Most of the forest was made up of young conifers. Even in sections where trees were densely packed, I reckon that trees were at least a couple of metres apart. Set out in a rectangular grid, that would mean a hectare of land (that's 100 metres square) might contain:

$$50 \times 50 = 2,500 \text{ trees.}$$

A square kilometre would have 250,000 trees, so four square kilometres – i.e. a square plot of just 2 km × 2 km – would give us about a million trees. That sounds a lot, but remember that a *trillion* is a million times bigger than one *million*.

Based on my crude estimate, to reach the target of one trillion trees we'd therefore need:

$$4 \times 1,000,000 = 4,000,000 \text{ square km}$$

of densely packed trees to meet the trillion target.

Let's compare that to somewhere we know.

The area of Wales is roughly 20,000 km². France is a bit more than 500,000 km². India is about three million km². So we might be talking about new forestry that would cover 20 Waleses, eight Frances or India-plus-a-bit. I'll leave you to judge whether that sounds like a lot.

There's another way of looking at one trillion. There are seven billion people on the planet, so that means we need:

$$1 \text{ trillion} \div 7 \text{ billion} \approx 100 \text{ more trees}$$

per person on the planet.

Next question: where are we going to plant them? And who will do the planting?

FERMI FOR THE FUN OF IT

We've seen plenty of examples where back-of-envelope calculations can have a practical benefit – whether it's checking out the viability of a business plan, understanding our impact on the environment, or challenging a politician's statistical claims.

But there's no need to stop there. There are many Fermi-style questions that are no more than flights of fancy, a serendipitous exploration for those with a curious mind. Some people enjoy Fermi questions as a form of mental exercise – or as a way of passing the time when waiting in a queue. I was drawn into this form of thinking from an early age – whenever we were sat waiting for an event to start (a play, a cricket match, it didn't matter), my dad would invariably ask the open question: 'I wonder how many people there are here today' or 'I wonder what they make in ticket revenue.'

So, in the spirit of idle curiosity and wanting to exercise skills in estimation, here are some final Fermi questions that have little practical benefit, and their own peculiar fascination.

HOW LONG TO COUNT TO A MILLION?

If you have children, you've probably had that joyful moment when they start to count and realise that they could, in theory, keep on counting for ever. How high could they get?

Try counting aloud at a 'normal' speed . . . one, two,

three, four . . . you are probably counting at about two numbers per second. That suggests you'll get to 100 in about a minute, 1,000 in ten minutes, and one million in 10,000 minutes, which is about seven days if you manage to stay awake. Well, except . . . it takes longer to say big numbers. Start counting from (say) 243,100: how far do you get in a minute? Instead of two numbers per second, you're now probably taking over two seconds per number – that's only a quarter of the speed.

In the count to a million, the vast majority of numbers will be long, multi-syllabled ones. So two seconds per number is a reasonable estimate, which makes two million seconds. Even without sleep, that's about 25 days (pretty much one month) to count every number up to a million.

If you're American, you might shave a little bit off that time, because you don't use the word 'and' in your spoken numbers. What a Brit calls 'One thousand two hundred and four', an American calls 'one thousand two hundred four'. I estimate that's a saving of about 5% of spoken time.[32]

This time-shaving helped Jeremy Harper of Birmingham, Alabama, who currently holds the world record for counting to the largest number. In the summer of 2007, Harper counted from one to one million. It took him just under three months. Given that he needed to sleep, eat and take mental breathers, it's impressive that this is only three times as long as our theoretical non-sleep minimum of one month.

Will anyone ever count higher than this? Well, there's one guy who's trying.

Count von Count, the Transylvanian Muppet who loves

---

[32] If you're French, 1,204 is the even punchier: *mille deux cents quatre*.

to count, has a Twitter account. Every day, the Count counts another number in words. Sometimes he counts two, or even three numbers in a day. Last time I looked, he was still in the low 2,000s. How long will it take him to get to one million? At this rate, roughly 500,000 days should do it, which is over a thousand years (500,000 ÷ 365 ≈ 1,000). But he doesn't have to stop there.

He could keep going until . . . well, until he exceeds his Twitter limit of 280 characters. But how long will that be?

Since the Count counts US-style, he doesn't have 'ands' to worry about. And let's assume that every time he adds another three zeroes, he moves up the standard scale, from billion to trillion to quadrillion.

The limit on how far the Count can go isn't the magnitude of the number. After all, the number 90,000,000,000,000,000,000,000,000,000 can be written in just 16 characters: *90 octillion!* (The Count always puts an exclamation mark at the end of his tweets.)

Before he gets to 90 octillion, however, he is going to encounter numbers that exceed his Twitter limit. For example, let's pick a random number that's over 20,000 times smaller than 90 octillion – 3,865,497,871,750,829,425,934,673. That requires 285 characters when written in full, Count-style. (In full, it's: *Three septillion eight hundred sixty-five sextillion four hundred ninety-seven quintillion eight hundred seventy-one quadrillion seven hundred fifty trillion eight hundred twenty-nine billion four hundred twenty-five million nine hundred thirty-four thousand six hundred seventy-three!*)

What's the *smallest* number that breaks the Twitter limit? To find it, we need to pack our number with the numbers that use up the most letters: seven and seventy. You might even want to see if you can work out exactly which number it will be before you look it up (hint – it's in the sextillions). Don't forget there are no commas, and the Count always ends with an exclamation mark. The answer is on page 184 at end of the book.

How long will it take the Count to get to this massive number? At an average of two numbers per day, it will take over 50 sextillion days, around 100 quintillion years. As I mentioned earlier, the universe will most likely end in the next few billion years, so I think we can be confident that running out of Twitter characters is one problem that the Count is not going to encounter.

HOW OFTEN DOES A TEENAGER SAY 'LIKE' IN ONE YEAR?

Do you ever listen to what people say? I mean, what they actually say, a transcript of the exact words that they use. It's intriguing to do so, because a conversation that seems fluent is often packed with pauses, corrections and so-called filler words. It's the filler words that interest me most, because they are used so often: words like 'um', 'you know', 'er', 'basically' and – the *bête noire* of many people – 'like'.[33]

Like has been prevalent in English-speaking cultures for over 20 years, particularly among teenagers in the UK and the USA. In fact I would go so far as to claim that there are

---

[33] Linguistics experts call the word 'like' a *discourse marker*, and have found that it plays a more important role in speech than fillers such as 'um' and 'er'. See, for example, 'Placing like in telling stories', Jean Fox Tree, 2006.

some individuals who say the word *like* more often than they say any other word in the English language, ahead of the supposed top 10 : *the, of, and, to, a, in, is, you, are* and *for.*

So how often might a 'typical' chatty American teenager say the word *like* in a year?

To make an estimate we need some data – which means capturing some real conversation. One way to do this is to eavesdrop in a bus queue, where there is no shortage of material. Or you could do as I did, and look for a chatty vlog channel on YouTube. Here's one snippet of a conversation between two individuals:

> Person A: 'Have you noticed that [name] has three voices, she has her normal talking voice **like** 'hi guys how's it going?' and then she has her **like** pissed-off voice where she's **like** 'yo! – **like** – don't ever do that again' and then she has her cutie voice where it's **like** [puts on silly voice] a wittle wabbit.'
>
> Person B: 'I **like** her cutie voice.'

That snippet of a high-energy conversation lasted 15 seconds. Between them they spoke 63 words, and the word *like* appeared seven times. They didn't always use 'like' as a filler. Sometimes it was as a preposition ('like a wittle wabbit') and sometimes as a verb ('I like her cutie voice'). But when it comes to 'like' counting, they all count.

In the conversation snippet, the word 'like' represented more than 10% of the words spoken. I call this figure the 'like quotient'. A like quotient of 10% is not unusual – if you know somebody who says 'like' a lot, they probably use it at this rate. From my informal research, peak use of 'like' is around one in five words, or 20% – but even peak-likers don't usually keep up that rate in all their conversation. Meanwhile, there are many teenagers who barely use the like word at all.

So let's do the calculation for a chatty teenager with a 10% like quotient.

Let's assume:

- The average conversation rate is around 100 words per minute.
- The chattiest teenager in a group (whether that's two, three or more people) talks for about half the time. So the chatty teenager speaks 50 words per minute, which at a 10% like quotient (L.Q.) is 5 'likes' per minute.

How long does a chatty teenager spend in conversation in a day? Travelling to and from school each day, they may be engaged in conversation for, say, one hour in total. Let's add in another couple of hours in the day to cover lunch breaks and other social times. That's three hours – let's call it:

200 minutes per day × 5 likes per minute
= 1,000 likes per day.

1,000 × 365 days per year ≈ 400,000 likes per year.

That's, like, nearly half a million!

It wouldn't surprise me if peak-likers utter the word 'like' well over a million times per year.

Like, wow.

ARE YOU DESCENDED FROM RICHARD III?

In August 2012, a team of archaeologists digging in a city-centre car park in Leicester discovered a skeleton. DNA testing followed, which confirmed that they had found the long-lost body of Richard III, the king made into an infamous villain by Shakespeare as the hunchback monarch accused of murdering his young nephews in the Tower of London so that he could claim the crown. Richard of York (as he was also known) was defeated and killed by Henry Tudor at the Battle of Bosworth.

The euphoria at the rediscovery of Richard's bones was soon followed by an argument as to where they should be buried. The people of Leicester believed he should be buried in Leicester Cathedral, fittingly close to where he'd been found. But a small group, going by the name of the Plantagenet Alliance, declared that, as the descendants of Richard III, they should have the right to decide on the burial place of their ancestor. They wanted Richard to be buried in York.

I was struck by this story because, after 500 years, I expected that there might be rather more than 15 descendants interested in this case. So I dived into Google, investigated Richard's family, grabbed the back of an envelope and did some calculations.

Here's what I found.

Although Richard III is known to have had three children, his only legitimate heir died in childhood, and his two others – both illegitimate – are not believed to have had any offspring. So there are no known direct descendants of

Richard III. The so-called descendants lobbying on his behalf were actually descendants of his nephews and nieces.

Richard had five siblings, and a number of nephews/nieces (though Shakespeare reckons he killed two of them, the famous Princes in the Tower). Richard's eldest sister Anne had a daughter who had 11 children, and one of those grand-nieces herself had 11 children, so within a couple of generations there was a healthy stock of Richard III relatives to procreate.

Let's assume that each surviving descendant had two children who survived to child-rearing age, and that one generation is around 25 years. In the 500 or so years since Richard's death, there have been about 20 generations, so if there was no interbreeding between descendants, there are around 2^{20} descendants of his nieces and nephews today, which is about one million.

But that's a conservative estimate. If more than two children survive, e.g. 2.3 surviving children per generation (that's a low estimate for wealthy families, who had higher survival rates because of better nutrition), the number of descendants leaps to an incredible 17 million – an example of the 'sensitivity' I discussed on page 22.

However, we know that marriage between distant cousins is inevitable. As you come down through the descendants of Richard's family, there will be many who were marrying remote cousins. This will significantly reduce that figure of 17 million, but it's reasonable to suppose that there are at least a million people who are descendants of Richard III's siblings. To support this claim, Dr Andrew Millard of Durham University wrote a paper that makes a credible case for saying that if you have any English ancestry, you are almost certainly a descendant of King Edward III, Richard's great-great-great-great grandfather. If it's true for Edward III, then the odds for Richard III featuring in your ancestry must be quite high as well.

My conclusion? Since there's a chance that anyone with English ancestry is related to Richard III, the Plantagenet Alliance had no more right than the rest of us to decide on Richard's burial place. That was also the view of the three High Court Judges, who dismissed the case when it went for judicial review. It was a victory for back-of-envelope maths (though an expensive one: it cost taxpayers over £200,000[34]).

HOW MUCH FURTHER CAN YOU THROW A SHOT PUT WHEN YOU ARE IN MEXICO CITY?

You might think that it makes no difference where you are when you throw a shot put. (Well, if distance is the objective, I suppose the top of the Eiffel Tower would be good – but I'm assuming we're on a flat piece of ground here.)

It turns out, however, that there are two variables that can have a significant effect on how far a projectile will travel: air density (which affects air resistance) and gravity.

This was famously demonstrated in 1971, when the *Apollo 14* Commander Alan Shepard pulled off a memorable stunt while standing on the moon. The moon has about one-sixth of earth's gravity, and almost zero air resistance. Shepard

---

[34] It reportedly cost the Ministry of Justice £90,000 in admin and Leicester Council £85,000 in legal fees.

had somehow managed to smuggle the head of a golf club and a couple of golf balls onto the spacecraft. On the moon, he improvised the handle of a golf club, and with a one-armed swing he managed to whack one of the balls – in his words – 'miles and miles'. Later, he gave a more realistic assessment that the ball had gone 'over 200 yards'. With a proper club and two-armed swing, it's reckoned an astronaut on the moon could easily hit a ball over a mile.

Back on earth we can't avoid air resistance, but the density of the atmosphere does change with altitude. There is less air resistance at the top of Table Mountain than there is at the Dead Sea. So – other things being equal – a shot put released on top of a flat mountain will travel a little further horizontally than one released at the same speed at ground level. The difference, however, is small. Air resistance has a big effect in slowing a balloon, but on a dense lump of iron its impact is minimal. Even in a vacuum, a shot put would travel only a couple of centimetres further than it would in the open air.

Gravity, however, is a different matter. Although we were all taught in school that gravity is a 'constant' on the surface of the earth, this is not strictly true. There are two things that affect it. The first is that gravity reduces the further you go from the centre of the earth. So (like air density), gravity is smaller on a mountain top than at sea level.

The second is that while gravity is pulling us in to the centre of the earth, there is something else that is attempting to fling us outwards. Like a roundabout in the playground, the earth is spinning around, and without gravity we would be flung out into space. The faster you spin, the greater the force. Close to the North Pole, the speed at which you are spinning around the earth's axis is close to zero, but at the equator you are hurtling around at about 1,000 miles per hour.

These two factors – the lower gravity due to height, and the reduced gravity due to centrifugal force – mean that gravity is detectably lower at altitude on the equator (in Mexico City, for example) compared to sea level near the North Pole (Helsinki, say). In Mexico, gravitational acceleration (usually denoted as 'g') is about 9.77 m/s², while in Helsinki it is 9.83 m/s². These figures do vary, but that's a difference approaching 1%.

What impact does this have on the range of a shot put?

There is a formula for the range of a projectile, based on Newtonian physics. I was going to show it here, but my editor warned me that it might immediately knock about 20% off the sales of this book. So I've buried the full thing in the Appendix (page 183) and simplified it to this:

$$\textit{Distance travelled} = \frac{k}{\sqrt{g}}$$ (K is a constant, g is acceleration due to gravity).

OK, even that might not look 'simple'. What it means is that as the value of g decreases, the distance that the shot put travels increases (by an amount proportional to 1 divided by the square root of g).

A good shot-put throw travels about 20 metres. If we reduce gravity by 1%, we increase the range of the shot put by about ½%, or 10 cm. That's not to be sniffed at when world records are at stake.

WHERE ARE THE ALIENS?

Aside from his general approach to back-of-envelope questions, Enrico Fermi is remembered for one calculation in particular.

Sometime not long after the end of the Second World War, Fermi and some other scientists were having a conversation, and the subject turned to extraterrestrial beings. As the story goes, Fermi suddenly asked the question: 'So, where is everybody?' – by which he meant, given the billions of stars in the galaxy, one of which must surely have developed advanced life forms, why hadn't we been invaded by aliens yet?

His question became known as the *Fermi Paradox*.

Several years later, the astrophysicist Frank Drake came up with an equation for expressing the number (N) of intelligent, communicating civilisations that are in our galaxy at any one time. His equation was this:

$$N = R^* \times n_p \times f_L \times f_i \times f_c \times L,$$

where:

R* is the average number of stars that form each year;
np is the average number of planets per star;
fL is the fraction of those planets that develop life;
fi is the fraction of planets with life that develop *intelligent* life;
fc is the fraction of civilisations that develop communication technology;
L is the number of years that communicating civilisations survive.

Although it looks sophisticated, this is really just common sense set down in mathematical form. The hard part – and the estimation – comes in putting values to each of the factors.

For example, of the planets that form around any particular star, what proportion are capable of developing life? To even have a stab at coming up with a sensible figure requires an understanding of the chemicals and physical environment that are essential for life forms to be able to exist.

Various scientists have attempted to come up with sensible figures for each of the factors.

For the average number of stars that form each year, suggestions have ranged from 1 to 10. The number of planets per star that can develop life is reckoned to be in the range of 0.2 to 2.5, of which the percentage that form intelligent life is anywhere between 1% and 10%, and of these anything between 1% and 100% develop the technology to communicate. When they do communicate, different scientists have reckoned the civilisation will last anywhere between 100 and one billion years (with one scientist coming up with the suspiciously precise figure of 304).

Picking middling values for each of the factors gives us:

$$3 \times 0.5 \times 10\% \times 25\% \times 30\% \times 1000 \approx 10.$$

So that suggests there might be 10 civilisations out there that are capable of communicating, and that might therefore be detectable.

However, this number is extremely sensitive to the figures that you plug into the equation. The estimates for the number of civilisations out there communicating with the galaxy at any time could be anywhere between 1×10^{-10} (which is effectively zero) and 15 million. This must be the widest range of answers for a back-of-envelope question that has ever been produced, and it makes the forecasts for vCJD described on page 26 look like nanotechnology precision.

The Drake equation is a fun intellectual exercise, but it's also a good place to stop, because it demonstrates that, sometimes, attempts at estimation are bordering on the futile.

WHEN THE ROBOTS TAKE OVER

THE *COUNTDOWN* CONUNDRUM

Before we finish, let's briefly move away from estimation and back to the world of precise arithmetic.

In a classic 1997 episode of Channel 4's word and numbers game *Countdown*, the host, Carol Vorderman, picked the following six numbers from the table in front of her (four from the top, and two from the third row):

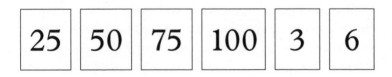

The random number generator then produced a target of 952. The challenge, as always, was for the contestants to use some or all of the numbers on the cards no more than once, to get as close as possible to the target answer of 952.

Perhaps you'd like to have a go now, to see how close you can get.

To get close to the target answer, it's a huge help to be able to play with numbers. And curiously, even though the task involves 'exact' calculation, rough estimates can be handy as a starting point: '952 . . . that's going to be 9 × 100 and a bit . . . or 1,000 minus 50ish'.

If you managed to get 950 (two away from the target), give yourself a bronze medal. This is the way that most people get there:

$$100 \times (3 + 6) + 50 = 950.$$

If you managed to get to within one (953), give yourself a silver medal. To create the 3, you need to spot that:

$$75 \div 25 = 3$$

$$950 + 3 = 953.$$

That would normally be enough to win you the round, but on the programme in question, you would have lost out to the other contestant, James Martin, a PhD maths student who managed to get the exact answer of 952.

Here's a slightly shortened version of his exchange with Carol Vorderman as he explained how he'd got his answer:

JM: 100 + 6 = 106. Multiplied by 3 . . .
CV: . . . is 318.
JM: I'd like to multiply it by 75 . . .
CV: Multiply 318 by 75? [Laughter] Good grief, I'm going to need my calculator for this one. [Eventually she does the calculation to get 23,850.]

JM: Now take away 50.

CV: [More laughter] 23,800.

JM: And divide it by 25.

CV: And divide THIS by 25? . . . [Can barely control
     hysterics as she writes out the division] Do you
     know – I think you're right. That's incredible!

There are many calculation savants who could do all of the calculations above – and harder ones – in their heads. But James Martin was not a savant, he was just smart at manipulating numbers.

Martin spotted that $106 \times 9 = 954$, which is 2 away from the target.

There wasn't a 9 available, but he could get to the same answer by multiplying by 3 twice: first by the 3 on the card, then by $75 \div 25$, which is also 3. But where would he get the 2 that he needed to subtract from 954? What James spotted was that $50 \div 25 = 2$, so he could use the 25 twice, by dividing it into 75 and 50.

Written out in full, his solution was this:

$$\frac{((100 + 6) \times 3 \times 75) - 50}{25}$$

This might make it appear that he multiplied 318×75 (as you'd do on a calculator), but before multiplying, he divided 75 by 25 first to get 3 and 50 by 2 to get 2. This turned the calculation into:

$$(106 \times 9) - 2.$$

Which is clever. But it is not genius.

Back in 1997, when that solution was recorded, it's unlikely that even somebody armed with a computer could have beaten James Martin to the score. But today, there are apps that can solve these *Countdown* problems instantly. It won't be too long before a contestant could be wearing a pair of glasses that detects the numbers on the board and displays the solution on the lenses before the timer has even begun to count down.

Which leads to an interesting *Countdown* conundrum.

We are not far away from a world where we will all be equipped with Artificial Intelligence devices that can solve any numerical puzzles of this kind in an instant. When technology like this becomes readily available, not only will we not need calculators – some might question why we will need to learn any maths at all. When robots can work everything out, will games like *Countdown* continue to exist?

There will of course be people who scoff at anybody who 'wastes their time' figuring out number puzzles when the solution is readily available to them. Just as there are many who scoff at the need to know how to do short division calculations manually, when it can be done by a calculator.

But my prediction is that in 50 years' time – even when computers can solve just about any numerical problem posed to them almost instantly – people will still have a huge interest in playing number games in their heads. And it won't just be for a bit of TV amusement. We *need* to continue to be able to do back-of-envelope calculations without the aid of a calculator, or other artificial device.

Why?

Because we will always need to be equipped to challenge the information that is presented to us, whether it has come from a person or from a computer. If we leave every calculation and every decision to computers, we are in danger of becoming slaves to technology.

And aside from all the practical benefits of being able to do maths on the back of an envelope, there is arguably another that is just as important: doing your own calculations keeps the brain stimulated, and gives it a valuable work-out. Some of us go further, and regard it as fun.

APPENDIX

SIGNIFICANT FIGURES

The idea of rounding a number to one, two or three significant figures is a recurring theme in this book. Here's a reminder of how it's done.

Let's take the height of the Matterhorn in the Alps, which according to most sources is 4,478 metres. It's likely that surveyors have readings that give them a height to the nearest centimetre, but given that mountains are always on the move, they have understandably rounded the Matterhorn's height to a four-digit number that is correct to the nearest metre. The statistic therefore has four significant figures.

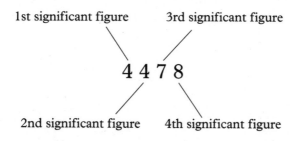

To round the number to three, two, or one significant figure, lop off the number at the end and replace it with a zero – but with one proviso: if the digit being removed is 5 or more, the digit before it should be rounded up by one.

Here's what happens when we round the Matterhorn height:

Rounded to . . .	
Three significant figures	4,480 (note that the 7 has been rounded up to 8)
Two significant figures	4,500 (the second 4 is rounded up to 5)
One significant figure	4,000 (note that 4,478 rounds down)

The first significant figure of a number is always its first non-zero digit. So, for example, the first significant figure of 0.0063 is 6. It's possible for a significant figure to be zero, including a number's final digit. For example, if an athlete runs 100 metres in 10.28 seconds, that is 10.3 seconds to three significant figures, and 10 seconds to two significant figures.

WHERE THE RULE OF 72 COMES FROM

The 'Rule of 72' is found by working out how many iterations (years, for example) it takes for a number to double if it is growing at a fixed rate. To follow the derivation, you do need to be familiar with natural logarithms.

Let's call the annual interest rate R per cent. What we are looking for is the number of years, N, that it will take our starting pot of money, A, to double; i.e. after N years we will have an amount 2A:

$$A \times (1 + R)^N = 2A$$

Cancelling A on both sides:

$$(1 + R)^N = 2$$

Take logarithms of both sides:

$$N.\ln(1 + R) = \ln 2 = 0.69 \, (= 69\%)$$

There is a rule of thumb familiar to mathematicians that if R is small then $\ln(1 + R) \approx R$ (this is accurate to within 5% if $R < 10\%$). In other words:

$$N \times R = 0.69$$
$$N = 69\% \div R.$$

This is why it should really be the Rule of 69. The number is adjusted to 72 because 72 is a multiple of many standard interest rates: 1%, 2%, 3%, 4%, 6%, 8% and so on.

WHO WANTS TO BE A MILLIONAIRE? (PART 2)

The ocean is the Arctic. John estimated that the Atlantic is about 30 million square miles, using similar calculations to the ones on page 89. That's getting on for 10 times bigger than the 4.7-metre-square-mile ocean in the question. The Indian Ocean is a similar size to the Atlantic, and the Pacific is bigger than both of them. No doubt the reason most of the audience voted for the Pacific is that 4.7 million is a really big number, and they knew that the Pacific is also really big. But, of course, 'really big' and 'really *really* big' are not the same thing.

RANGE OF A SHOT PUT

The range of a shot put can be worked out from this complex-looking formula:

$$Rd = \frac{v^2}{2g}\left(1 + \sqrt{1 + \frac{2gy_0}{v^2 \sin^2\theta}}\right) \sin 2\theta,$$

where:

R is the range of the shot put;

v is the speed of the shot put when released;

g is the acceleration due to gravity;

θ is the angle relative to the horizontal of the shot put when it is released;

y_0 is the height above the ground at which the shot put is released;

If gravity is the only thing that changes, then the range increases roughly as I indicated on page 168. In reality, in lower gravity it should be possible to release the shot at a higher speed (the shot will feel less heavy, so you can push it faster). This will *increase* the range, so my estimate of the advantage of Mexico City is on the low side.

HOW LONG TO COUNT TO A MILLION?

'Seven' and 'Eight' are the longest words from 0 to 9, and 'seventy' is the longest tens word, so the smallest number to exceed 280 characters is going to depend heavily on sevens. 777,777,777,777,777,777 (seven hundred seventy-seven quintillion seven hundred seventy-seven quadrillion seven hundred seventy-seven trillion seven hundred seventy-seven billion seven hundred seventy-seven million seven hundred seventy-seven thousand seven hundred seventy-seven!) uses 255 characters. We have 25 characters left. In front of that number we put the sextillions (sextillion has 11 characters including the space). The smallest number to exceed 25 characters is: one hundred one sextillion. So the Count will be frustrated when he gets to 101,777,777,777,777,777,777.

Phew!

ANSWERS AND TIPS

ENVELOPES VERSUS CALCULATORS (PAGE 8)

(a) $17 + 8 = 25$ The vast majority of adults and teenagers in my survey answered this in their head, but even for a straightforward addition like this, there is a range of ways that people use to get there. The three common approaches were:

- $7 + 8 = 15$, then add 10 to get 25.
- Split 8 into $3 + 5$; then add $17 + 3 = 20$; then add 5 to get 25 (splitting a number up in this way is referred to as partitioning in primary schools).
- 8 is two less than 10 . . . $17 + 10 = 27$, then take away 2 to get 25.

(b) $62 - 13 = 49$ Almost everyone uses two steps here. Those who do it mentally either do 'take away $10 = 52$, then take away three $= 49$' or 'take away three $= 59$, then take away $10 = 49$'. Those who use a written method will typically work from the right: '3 from 2 . . . borrow 10 . . . , etc.

(c) $2,020 - 1,998 = 22$ Viewed as a regular subtraction, $2,020 - 1,998$ requires careful carrying over of tens and hundreds. But if the problem had been: 'Amy was born in 1998. How old will she be in 2020?' most people solve this by counting up, rather than by subtracting: 'two years from 1998 to 2000, add 20 years up to 2020, equals 22 years'.

(d) $4 \times 9 = 36$ Those who do calculations frequently will remember their times tables, and regurgitate 'four nines are thirty-six' without having to think. But it's interesting to observe how those who are rusty with their times tables work it out. The quickest way is to calculate 10×4 (= 40) and then subtract 4.

(e) $8 \times 7 = 56$ Aside from instant recall, approaches that adults shared with me included:
- $7 \times 7 = 49$, add 7 = 56
- $2 \times 7 = 14$, double it = 28, double that = 56
- $5 \times 7 = 35$, add 7, add 7, add 7 = 56.

(f) $40 \times 30 = 1,200$ People generally know, or very quickly work out, $4 \times 3 = 12$. But change it to 40×30, and the addition of those zeroes can make this calculation much more of a struggle. A common approach is to do it in two steps: reduce one of the numbers to a single digit (e.g. $40 \times 3 = 120$), then multiply by 10 to give $40 \times 30 = 1,200$. There are others, however, who just guess. The answer 12,000 is not uncommon. (See also the 'count the zeroes' rule on page 52.)

(g) $3.2 \times 5 = 16$ Of the calculations so far in the quiz, this is the first for which almost everybody uses a written method. Most common is: $5 \times 3 = 15$; $5 \times 0.2 = 1$; $15 + 1 = 16$.

Some shortcuts for multiplying by 5 can be found on page 42.

(h) $120 \div 4 = 30$ There are two common strategies for dividing by four. The first is to halve the number and then halve it again ($120 \div 2 = 60$, and $60 \div 2 = 30$). The other is to do mental short division: 'four into 12 goes 3, so the answer is 30'.

(i) Three-quarters   75% is such a commonly used quantity that many people are familiar that it is three-quarters without needing any thought. Some figure it out by starting at 25% (one-quarter) and multiplying it by three.

(j) 10% of 94 = 9.4   The most common approach that adults use is to shift the decimal place one to the left, so that the hundreds become tens, tens become units etc.

ARE YOU AN ARITHMETICIAN? (PAGE 37)

(a) £2.77 change. This sort of problem would usually be regarded as a subtraction, yet somebody working behind a bar would typically treat it as an addition: start with £7.23, round up 7p to £7.30, then add 70p to get £8, then add £2 to get to £10.

(b) Gandhi was 78 when he died. Done as a subtraction (1948 take away 1869) this can get messy, even before getting your head around the months of Gandhi's birth and death. As with the change example in (a), it's easier if you treat this as an addition: It's 31 years from 1869 to 1900, plus 48 years after 1900. 31 + 48 = 79. However, Gandhi died before his October birthday, so he was still 78 years old.

(c) 56,000 pence or £560. The most common error here is getting the number of zeroes wrong. You'll find tips on placing the zeroes and decimal point on page 53.

(d) Her new salary is £28,840. In surveys of adults, typically only around 25% (that's one in four) are comfortable doing calculations involving percentages – even when they have a calculator.

(e) 32 miles per gallon. Dividing 144 by 4.5 would be beyond the mental arithmetic capabilities of almost anybody. The 'trick' here is knowing how to make the calculation easier. 4.5 is a difficult number – nobody wants to divide by that. But if you double 4.5 you get 9. Instead of 144 ÷ 4.5, you can double top and bottom to turn it into 288 ÷ 9. With mental short division (see page 43), this is then relatively easy: 9 into 28 goes 3, remainder 11, 9 into 18 goes 2; answer . . . 32.

(f) £28.80.

(g) 16% of 25 is 4. Most numerate adults confronted with this question break it down into steps: to find 16%, first work out 10%, then 5%, then 1%. Which is fine – it works. But it can be turned into a single step if you realise that 16% of 25 is exactly the same as 25% of 16.

(h) 54%. We're now reaching the calculations where almost everyone would automatically reach for a calculator. How do you even begin to figure this answer out exactly? Forced to do it mentally, some people spot that 38 ÷ 70 is close to 40 ÷ 70, which is four-sevenths. If you know that ¹/₇ is roughly 14% (and quite a few people do) then ⁴/₇ is going to be four times that: about 56%. But now to adjust this down – a bit of intelligent tweaking suggests it's going to be somewhere between 54% and 55%, but which is it? Mental short division (see page 43) delivers an answer in a few seconds: it's 54.3%, or 54% to the nearest whole per cent.

(i) 6,102. If you attempt 678 × 9 using long multiplication you'll probably get in a real tangle, as you try to carry figures over in your head. The short cut here (which you can use whenever multiplying by 9) is to multiply by 10 instead. 678 × 10 = 6,780. Now subtract 678. Answer – 6,102.

(j) 900. Working out the square root of 810,005 precisely is hard – that '5' on the end is a pain. But the square root of 810,000 is much easier. 9^2 is 81, and 900^2 is 810,000. You'll find a general method for working out square roots on page 86.

MULTIPLICATION AND TIMES TABLES (PAGE 42)

Here are some ways that you might have got to the answer:

(a) $3 \times 20 = 60$, plus $3 \times 6 = 18$, giving the answer 78. Or double 26 (= 52) then add 26 (= 78).

(b) $35 \times 10 = 350$, subtract 35 to get 315.

(c) Multiplying by 4 is the same as doubling twice: $171 \times 2 = 342$, and $342 \times 2 = 684$.

(d) Multiplying by 5 is the same as dividing by 2 and multiplying by 10, hence $462 \div 2 = 231$, and $\times 10 = 2{,}310$.

(e) Dividing by 5 is the same as multiplying by 2 and dividing by 10: $1414 \times 2 = 2{,}828$, divide by 10 to get 282.8.

MULTIPLYING FRACTIONS (PAGE 47)

(a) $\frac{1}{3} \times \frac{1}{2} = \frac{1}{6}$.

(b) $\frac{2}{5} \times \frac{2}{3} = \frac{4}{15}$, just over one-quarter.

(c) $\frac{3}{4} \times \frac{1}{5} \times \frac{2}{3} = \frac{6}{60} = \frac{1}{10}$. (You can cancel out the 3s on the top and bottom to simplify the calculation to $\frac{1}{4} \times \frac{1}{5} \times 2 = \frac{2}{20}$.)

(d) $\frac{6}{7} \times \frac{14}{23} = 6 \times \frac{2}{23} = \frac{12}{23}$, i.e. just over one-half.

(e) To work out $\frac{51}{52} \times \frac{50}{51}$ cancel the two 51s to give you $\frac{50}{52}$ (which is about 96%). Here's a real-world application of this calculation: the chance that the Ace of Spades won't be the top card *or* the second card in a shuffled pack of 52 cards is $\frac{51}{52} \times \frac{50}{51}$, which is $\frac{50}{52}$.

PERCENTAGES (PAGE 50)

(a) £21 (25% of £28 is £7).

(b) 12 (10% of 80 is 8, plus 5% (4) = 12).

(c) 7. (Remember that 14% of 50 is the same as 50% of 14.)

(d) About 70%. (49 ÷ 68 is close to 49 ÷ 70, and 4.9 ÷ 7 = 0.7.)

(e) 44%. (Using short division, 2.66 ÷ 6 = 0.44 . . . and you can stop there.)

(f) New salary £27,100. A short cut here is to spot that 8.4% of 25 is the same as 25% (or one-quarter) of 8.4 = 2.1. So Kate's salary increase is 2.1 × 1,000 = £2,100. (You'd get a good estimate of the answer by saying that 8.4% is roughly 10%, so her pay rise will be a bit below £2,500.)

MULTIPLICATION (PAGE 53)

(a) 36,000 (4 × 9 = 36, and three zeroes).

(b) Four zeroes, so 210000 = 210,000.

(c) 88 followed by six zeroes, so 88 million.

(d) 50 × 50,000 = 25 followed by five zeroes, 2,500,000 – which was one-tenth of their target. (A true story, by the way.)

DIVIDING BY LARGE NUMBERS (PAGE 55)

(a) $100 \div 2 = 50$.

(b) $630 \div 9 = 70$.

(c) 2,000,000 (2 million).

(d) The same as $220 \times 0.3 = 22 \times 3 = 66$.

(e) The same as $50 \div 1 = 50$.

USING STANDARD FORM FOR LARGE NUMBERS (PAGE 56)

(a) 40,000,000 (40 million).

(b) 1.27×10^3.

(c) $6 \times 10^9 = 6,000,000,000$.

(d) 2.4×10^{11}.

(e) 0.5×10^5 or, more correctly, 5×10^4.

(f) 3.5×10^7 (which is 35 million).

KEY FACTS (PAGE 60)

(a) A bit under 12,000 miles. New Zealand isn't quite halfway around the planet from the UK.

(b) About 3,500 miles. It's roughly quarter of the way around the earth. Or, if you have ever flown to New York, you will know that the flight takes six or seven hours. The plane will be flying at a bit below 600 mph. So the distance is going to be $600 \times 6 = 3,600$ miles (ish).

(c) About 9 million. Mexico is one of the world's bigger cities. So is London. Mexico's population is going to be nearer ten million than one million.

(d) 200 feet or 60 metres (though it will vary). If we allow 10 feet (3 metres) for each storey, then 200 feet/60 metres is a reasonable estimate.

(e) Three hours? Walking at a brisk 4 mph it would take 2½ hours, but few people can keep up that pace for long distances. Call it three hours.

(f) About five million. Children attend primary school between the ages of 4 and 11 (there are seven year groups, including reception). The very crude sketch below shows the spread of population in more economically developed countries. It's fairly even across all age groups from 0 to 70, and starts tailing off after that.

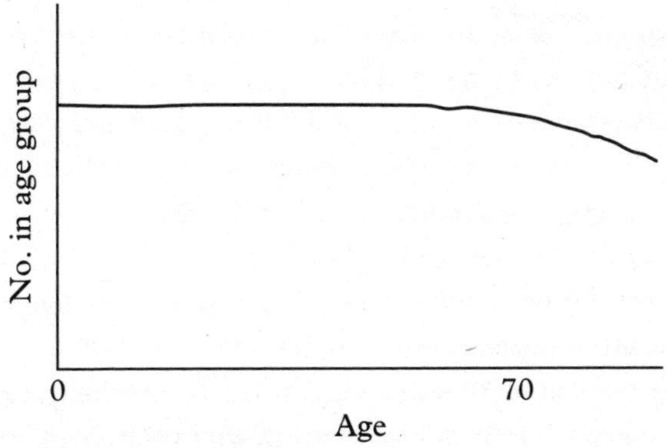

If we simplify this even more and imagine the population is evenly distributed across the ages 0 to 80, then in the UK there are roughly:

70 million people ÷ 80 years

~ 900,000 people in each age group.

That suggests around $7 \times 900,000 = 6.3$ million primary children – six million is a decent rounded estimate (the official figure is around 5 million).

(g) Around 200,000. From the previous question, we've got an estimate of around 900,000 people in each age group. Let's pick 30-year-olds and pretend that's the age when all people get married. If, say, half of all the population gets married at some point in their life, that makes 450,000 people getting married, and since it takes two to make a marriage, that suggests 225,000 (call it 200,000) weddings. Of course some people get married at 16, others at 60, but this doesn't change the answer, as long as people only get married once. In reality, some people do have more than one wedding, but this is the minority, so the average is unlikely to be more than, say, 1.2 weddings per person. That suggests around 250,000 weddings per year. (This isn't far from the official statistics, though the number of weddings is declining.)

(h) Official figures say anything between 30 and 40 million square miles. The Atlantic is a complicated shape. If we want to estimate its size, it's easier to think of it as a rectangle that fills the gap between Europe/Africa and the Americas. Let's call the width of the rectangle 3,500 miles (the London–New York distance); see answer (a) above. The Atlantic spans a large part of the globe north/south, so call its height 10,000 miles. The area is therefore roughly $3,500 \times 10,000 = 35$ million square miles.

ZEQUALS (PAGE 63)

(a) 83 $\approx$ 80.
(b) 751 $\approx$ 800.
(c) 0.46 $\approx$ 0.5.
(d) 2,947 $\approx$ 3,000.
(e) 1 $\approx$ 1.
(f) 9,477,777 $\approx$ 9,000,000.

CALCULATING WITH ZEQUALS (PAGE 65)

(a) 7.3 + 2.8 $\approx$ 7 + 3 = 10.
(b) 332 − 142 $\approx$ 300 − 100 = 200.
(c) 6.6 × 3.3 $\approx$ 7 × 3 $\approx$ 20.
(d) 47 × 1.9 $\approx$ 50 × 2 = 100.
(e) 98 ÷ 5.3 $\approx$ 100 ÷ 5 = 20.
(f) 17.3 ÷ 4.1 $\approx$ 20 ÷ 4 = 5.

INACCURACY OF ZEQUALS (PAGE 66)

(a) The biggest over-estimate is for 15.1 × 15.1 = 228.01; according to Zequals it is 20 × 20 = 400, which is 75% higher than the correct answer.

(b) The most extreme under-estimate is 14.9 × 14.9 = 222.01. Zequals rounds this to 10 × 10 = 100, which is over 55% too low.

SHOPPING BILLS AND SPREADSHEETS (PAGE 71)

The total is short by £190.10 – the value at the top of the column has been missed out. If you start by adding the hundreds column, you get £600 . . . which looks promising. But using Zequals, for example, you end up at £800, more than £100 above the £697.36 total, which should raise your suspicions. Rounding all the numbers to the nearest hundred also gives you £800. This is enough evidence to suggest that something is up – which indeed it is.

AREAS AND SQUARE ROOTS (PAGE 88)

These answers are correct to 3 significant figures. How close did you get?

(a) 5.10. If you estimated 5, give yourself a point. If you thought 5.1 – two points!

(b) 82.9. Split the number as 68 72. 68 is a bit more than 64, which is 8^2, so the answer will be a bit more than $8 \times 10 = 80$. If you estimated anything in the range 82 to 84, give yourself two points.

(c) 21.8. The decimal point is a distraction. 473.86 ≈ 500, so the answer is going to be a little more than 20. If you estimated 22, give yourself two points.

(d) 30.2. Split the number up as 9 10. This is close to 9 00, so the answer will be a bit more than $3 \times 10 = 30$, i.e. the room would be roughly 30 foot square (or 10 metres × 10 metres).

(e) 609 km × 609 km (roughly 400 miles square, so it would comfortably fit inside France). Split the number as 37 10 10. 37 is close to 36, so the answer is going to be a little more than 6 × 10 × 10 = 600.

BACK-OF-ENVELOPE CONVERSIONS (PAGE 95)

(a) Rough: 140 km   Accurate: 112 km
(b) Rough: 80 lb   Accurate: 90 lb
(c) Rough: 150 yards   Accurate: 163 yards
(d) Rough: 50 miles   Accurate: 63 miles
(e) Rough: 80 °F   Accurate: 77 °F
(f) Rough: 70 kg   Accurate: 62 kg

ACKNOWLEDGEMENTS

This book has been a long time in the making. I first contemplated writing it several years ago, yet it took a coffee with Wendy Jones to finally push me into action, and I'd like to thank her for being a great sounding board in the early stages.

My thanks to Hugh Hunt and John Haigh with whom I regularly have back-of-envelope jousts and who inspired a couple of the examples in the book. Graham Cannings, Chris Healey, Chas Bullock and Andrew Robinson provided invaluable critiques of the first draft, while Rose Davidson, Geoff Eastaway, Pete Sanders and Rachel Reeves provided equally helpful comments on the second. Special thanks to my mentor Dennis Sherwood for helping me to see the big picture and for sharing your deep insights on precision and the abuse of statistics.

I was lucky to be able to tap into some valuable expertise when I needed it, particularly from Claire Milne, Aoife Hunt, Jay Nagley and Ian Sweetenham, and also from Tom Rainbow and Catherine van Saarloos who are doing great work promoting *Core Maths* in schools.

My wonderful wife Elaine has been ever-present and ever-patient when I've sought advice on everything from grammar to subtitles.

Thank you Timandra Harkness for reminding me about the cats.

And finally, thanks to the team at HarperCollins, in particular Ed Faulkner who so enthusiastically picked up on this book, and to my lovely editor Holly Blood, who has so professionally filled the dual role of constructive critic and supportive cheerleader.

ABOUT THE AUTHOR

Rob Eastaway has written or co-written twelve books, including the bestselling *Why Do Buses Come in Threes?* and *Maths for Mums and Dads* as well as *What Is a Googly?*, the acclaimed beginner's guide to cricket. He is the Director of Maths Inspiration, a national programme of theatre-based lecture shows for 15–17 year olds that has reached over 150,000 teenagers since it began in 2004.

Rob regularly gives talks to all age groups in primary and secondary schools, appears on BBC Radio 4's current affairs/numbers programme *More or Less*, and works closely with National Numeracy, the national charity that campaigns for better adult numeracy.

In 2017, Rob received the Zeeman medal for excellence in communication of maths to the general public.